Leading When Machines Think Faster

Winning with AI: The Leadership Playbook for Enterprise Advantage

By

Chandan Lal Patary

From the Author of the books, The Scrum Master Guidebook, The Product Owner Guidebook, High-Performance Team Coaching Guidebook, The Innovation Blueprint and Master your Mind, Master your leadership

notionpress.com

INDIA • SINGAPORE • MALAYSIA

ISBN 979-8-90451-183-8

Also, by Chandan Lal Patary

1. *THE INNOVATION GUIDEBOOK*
2. *101 ENTERPRISE BUSINESS TRANSFORMATION CASE STUDIES*
3. *THE PRODUCT OWNER GUIDEBOOK: A PRAGMATIC REFERENCE MANUAL FOR MATURING PRODUCT COACHING*
4. *A GUIDEBOOK OF COACHING HIGH-PERFORMANCE TEAM*
5. *WE CAN LEAD: A GUIDEBOOK OF PERSONAL LEADERSHIP AND SELF-COACHING*
6. *THE SCRUM MASTER GUIDEBOOK: A REFERENCE FOR OBTAINING MASTERY*
7. *THE AGILIST'S GUIDEBOOK – A REFERENCE FOR AGILE TRANSFORMATION*
8. *50 TOOLS TO COACH YOUR WAY TO SUCCESS*
9. *MASTER YOUR MIND, MASTER YOUR LEADERSHIP*
10. *THE IMPACT COACHING GUIDEBOOK*
11. *CONTINUOUS TRANSFORMATION: THE SACRED WAY*

About the Author

Chandan Lal Patary: A Journey of Transformation, Knowledge, and Purpose

Chandan Lal Patary, based in Bangalore, Karnataka, India, is a seasoned transformation coach, technologist, and thought leader in enterprise agility, leadership, and innovation. He lives with his wife and two children and brings with him a diverse, 27+Years of long journey filled with continuous learning, impactful leadership, and knowledge sharing.

Chandan began his professional life as an apprentice engineer in an electrical machine repair company and later transitioned into the software industry as a trainer. Over time, he grew into roles such as test engineer, developer, technical lead, project manager, program manager, global program manager, engineering manager, and finally into his current role as an enterprise agile transformation coach. His career trajectory is a testament to his adaptability, resilience, and lifelong learning mindset.

He has accumulated a rich tapestry of experience across various industries such as Retail Fashion, Oil & Gas, Banking, Healthcare, Aerospace, Building Automation, Power Automation, Consumer Electronics, and Industrial Automation. His work has spanned companies like GE Medical Systems, Honeywell, ABB, Société Générale,

Royal Dutch Shell, Samsung, and H&M. He has collaborated with global teams from the USA, Germany, Sweden, China, Australia, Finland, Switzerland, France, Poland, Korea, London, and the Netherlands—experiences that have sharpened his technical acumen and cross-cultural leadership skills.

Chandan holds a Bachelor of Engineering in Electrical Engineering from NIT Agartala (1998) and an Executive General Management Program certification from IIM Bangalore (2007). He is a certified PMP (2008), Green Belt holder (2005), Certified Scrum Master (2011), and SAFe Agilist (2017).

One of Chandan's most profound experiences was in product development. While working with GE Datex Ohmeda in Finland, he co-developed patient monitoring systems for critical care units, gaining deep insights into real-time systems and software architecture. Later, with Honeywell Aerospace, he helped deliver aviation software for cockpit systems used by Airbus and Boeing, including taking part in flight tests in Seattle—an unforgettable highlight of his engineering career.

He then joined ABB Power Automation in Sweden, where he worked on mission-critical industrial control systems. This role provided him with deep insights into distributed Agile software development. His other key roles have spanned from managing large banking applications at Société Générale, oil & gas solutions at Shell, to driving omnichannel retail transformation at H&M.

Chandan stumbled into coaching in 2012 when a Head of Global Product Management in Finland recommended him for the role. What began as a suggestion turned into his life's mission: to facilitate enterprise-wide transformation through agility and leadership development.

As a transformation agent, Chandan has dedicated over a decade to studying Organizational Development, Business Agility, and the intersection of innovation, strategy, and people leadership. His work focuses on how these elements can accelerate organizational growth, People leadership development, and create meaningful, lasting change.

Chandan is also a prolific author and educator. He has authored several best-selling books including:

1. ***The Agilist Guidebook – A Reference for Organizational Agile Transformation (2018),***
2. ***The Scrum Master Guidebook – A Reference for Obtaining Mastery (2019),***
3. ***We Can Lead – A Guidebook of Personal Leadership and Self-Coaching (2020),***
4. ***A Guidebook of Coaching High-performance team (2021),***
5. ***The Product Owner Guidebook (2022),***
6. ***101 Enterprise Business Transformation Case Studies (2023).***
7. ***Business Metamorphosis: 50 Tools to Coach Your Way to Success (2024)***
8. ***The Innovation Blueprint (2024)***
9. ***and Master Your Mind, Master Your Leadership (2025)***
10. ***The IMPACT coaching Guidebook (2025)***
11. ***Continuous Transformation: The SACRED Way (2025)***

Beyond books, Chandan has contributed over **10,000+** blogs on LinkedIn, authored seven free eBooks on SlideShare, and published more than 50 technical papers across domestic and international journals, including 21 on DZone. He has delivered 20+ talks at conferences and created 30+ public presentations. In recognition of his work, he received the PM World Journal Editor's Choice Award in 2017 for his paper on business agility through organizational transformation.

His ultimate goal? To build a **"Body of Knowledge (BoK)"** that can empower future leaders and change agents. His mission is rooted in sharing real-world discoveries, encouraging collaborative problem-solving, and creating guidebooks and tools that others can leverage.

Purpose: Share the research and practical experience I have gathered to help others overcome transformation challenges and build organizational strength.

Vision: To create a globally recognized knowledge ecosystem that supports enterprise transformation and leadership development.

Mission: To continuously document insights and results from real-world challenges and disseminate them through practical, accessible formats.

Join Chandan in his journey of knowledge, growth, and transformation. Learn more at: **https://chandanpatary.com**

Acknowledgments

The contributions of numerous people have influenced this book. I want to thank everyone who offered feedback, shared tales, or offered suggestions. This book was inspired by all my friends and coworkers from both my current and past businesses. In order to write this book, I would like to thank all the Scrum masters, Agile coaches, AI Technical Managers and and leaders with whom I have had contact or whom I have interviewed. I would like to thank everyone of my fans and readers on social media for leaving me comments and suggestions so I can get better.

The wonderful people I have had the pleasure of getting to know and working with contributed to this book in a truly collaborative manner. It is an honor for me to work with such supportive coworkers. They offer motivation to write more effectively and pointers for studying.

The hundreds of team members and clients that I have had the privilege of working with and listening to, coaching, advising, and learning from have allowed me to advance in my proficiency.

I would like to thank all my colleagues with whom I have discussed ideas and confirmed my knowledge. I appreciate the constant support and criticism from all 36,000+ of my LinkedIn contacts.

My sincere gratitude goes out to all the prior supervisors and mentors who have molded, supported, and encouraged my professional progress over the past two decades.

I am profoundly grateful for the valuable leadership lessons I have learned from remarkable individuals throughout my journey. Each mentor and colleague have contributed uniquely to my growth, imparting insights that have shaped my leadership philosophy. From understanding the importance of empathy to navigating complex challenges with resilience, these lessons have been instrumental in my professional development. I owe a debt of gratitude to each of them for the opportunities they provided and the wisdom they shared.

A heartfelt expression of gratitude extends to my beloved wife, children, and parents, who have been unwavering pillars of support throughout my endeavors. Their encouragement, love, and understanding have played an indispensable role in my ability to overcome challenges and achieve milestones. My wife's unwavering support has been my anchor, providing the strength needed to navigate the complexities of leadership. My children's boundless enthusiasm has infused joy into my journey, reminding me of the significance of passion and purpose. Lastly, my parents' guidance and wisdom have been a constant source of inspiration, instilling in me the values that guide my leadership approach.

Completing this task was not an individual achievement but a collective effort, and I am immensely grateful for the profound impact my family has had on my personal and professional growth. Their presence has touched every aspect of my life, making this accomplishment a shared success.

Contents

Introduction

The Blind Spot No One Saw Coming

Sarah Chen had done everything right.

For more than twenty years, she built her company the hard way—route by route, client by client, decision by decision. She knew her business inside out. She could sense delays before dashboards showed them. She could identify weak points in operations just by listening to a regional manager's voice. Investors admired her instincts. Employees trusted her judgment. The board called her "battle tested."

In 2023, when the board pushed for AI-driven route optimization, Sarah didn't resist. She approved the budget, hired consultants, and asked IT to "take care of it." To her, AI was important—but not central. Leadership, she believed, was still about people, relationships, and experience. Technology was supported. Leadership was human.

Eighteen months later, her company lost nearly **$2 billion in market value**.

The competitor that overtook them wasn't bigger. It wasn't richer. It was led by someone younger, less experienced, and far less "seasoned." But that leader understood one thing Sarah didn't realize she was missing:

AI isn't a tool you install.

It's a capability you lead.

Sarah didn't fail because she ignored AI.

She failed because she treated it like infrastructure instead of intelligence.

She delegated it.

She distanced herself from it.

She never changed how *she* thought.

By the time the numbers made the problem visible, the gap was already irreversible. Decisions were slower. Judgments were weaker. Opportunities were missed—not because the data wasn't there, but because leadership couldn't absorb it fast enough.

When Sarah stepped down, the post-mortem wasn't about algorithms or vendors. It was about leadership. About assumptions that no longer hold. About a model that worked brilliantly in the past—but quietly collapsed in the present.

This book begins there—not to criticize Sarah, but to ask a harder question:

If a leader like her can miss this shift, what might you be missing, too?

The stories that follow are not about bad leaders.

They are about great leaders facing a world that changed faster than their thinking.

You'll meet executives whose success became their blind spot. Leaders kept awake at 3 AM by fears they never voiced. Organizations that invested millions in AI and still failed—because human capability didn't scale with technology.

You'll also discover something hopeful.

AI does not make leaders obsolete.

It makes **certain leadership capabilities obsolete**.

And capabilities—unlike age, title, or background—can be developed.

This book is about that development.

Not how to use AI tools.

But how to **think, decide, learn, and lead** when intelligence moves at machine speed, and consequences scale instantly.

If you are willing to examine your assumptions, question your habits, and evolve how you lead, this book will give you a path forward.

If not, the next Sarah Chen story may feel uncomfortably familiar.

Let's begin.

Why Leadership Must Evolve Before AI Does?

AI did not arrive quietly.

It entered organizations through dashboards glowing with confidence, models trained on vast amounts of data, and promises of speed, accuracy, and scale that no human system could match. Decisions that once took weeks now took seconds. Insights arrived before questions were fully formed. Automation spread from back offices into boardrooms.

And yet—something unexpected happened.

Despite record-breaking investment, talent, and technology, many organizations came to a halt. Pilots worked but never expanded. Trust declined. Adoption became just a show. Leaders felt both empowered and overwhelmed at the same time. Teams accelerated, but clarity faded. Data grew, but understanding lessened.

AI Agents Are Becoming More Prolific Across Tech Applications, Growing with a 45% CAGR Over Five Years

AI AGENTS MARKET BY TECHNOLOGY 2024–2030[1] ($ BILLIONS)

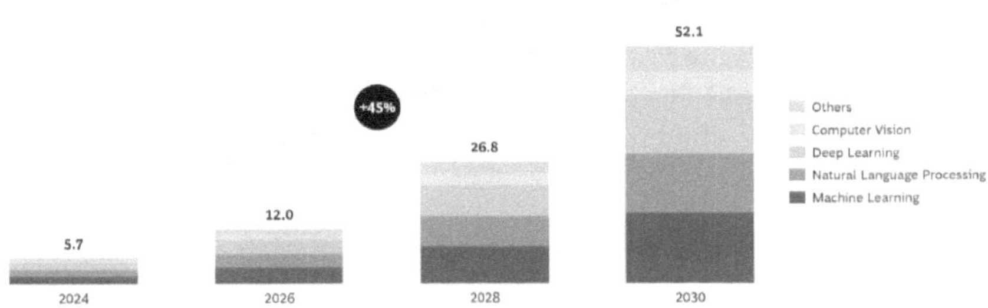

Source: BCG analysis.
Note: CAGR = compound annual growth rate
[1]GrandviewResearch.com

Generative AI Life Sciences, Education Spending

Generative AI Life Sciences, Education Spending

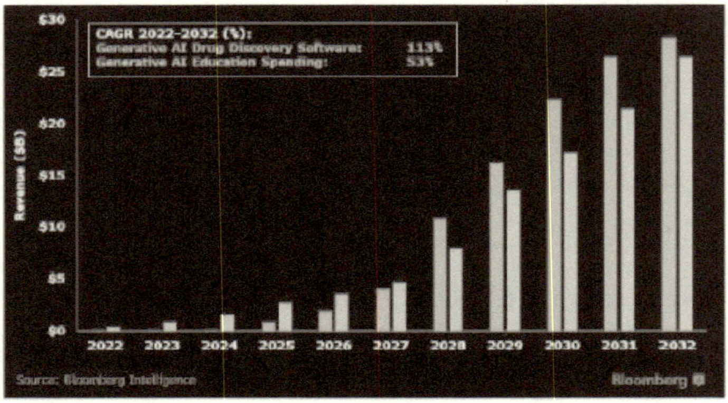

Source: Bloomberg Intelligence

The paradox became impossible to ignore:

AI was accelerating everything—except wisdom.

This book begins with a simple but uncomfortable truth:

AI is not breaking organizations.

It is revealing what was already fragile.

The Real AI Crisis Is Human

Most conversations about AI focus on models, platforms, architectures, and vendors. But after years of watching AI transformations succeed and fail across industries, one pattern repeats relentlessly:

AI initiatives fail far more often because of human and leadership gaps than because of technical ones.

Leaders are not unprepared because they lack intelligence or ambition. They are unprepared because the **pace and nature of AI change** demand capabilities that most leadership systems were never designed to develop.

AI creates pressures that traditional leadership models cannot absorb:

- Decisions multiply while consequences scale exponentially
- Expertise expires faster than identities can adapt

- Emotional load increases even as automation promises relief
- Ethics move from abstract values to daily operational choices
- Culture, trust, and judgment become performance constraints
- Human–AI collaboration becomes unavoidable, not optional

What once worked—confidence, control, expertise, linear planning—now quietly works against leaders.

The result is not failure by incompetence, but failure by **misalignment**.

Organizations try to install intelligent systems into leadership cultures built for a slower, simpler world.

And the system pushes back.

The Quiet Crisis Organizations Are Already Facing in the AI Era

Artificial Intelligence is moving faster than most organizations can absorb.

According to **McKinsey's Global AI Survey (2024)**, more than **70% of companies report adopting some form of AI**, yet only **about 20% say they are seeing significant business impact**.

The technology is advancing quickly.

But leadership systems are struggling to keep up.

This gap is already causing real damage inside organizations.

Problem 1 — Leadership Models Built for a Slower World

The traditional leadership models that support most modern organizations were created in a very different era—a time when information moved slowly, decisions went through lengthy, careful review, and deep experience was the main form of reliable guidance. In that setting, deliberation was a strength, and slow-moving hierarchies offered necessary stability.

Today, AI has completely changed those rules. Algorithms now produce complex insights in seconds, yet many organizations are still bound by layers of review and approval that can take weeks or even months. A report from the Deloitte AI Institute emphasizes how much this slows things down, noting that nearly 60% of executives admit their internal decision-making processes are too slow to fully utilize AI-driven insights.

This creates a structural mismatch: while our machines are speeding up toward real-time decision-making, our organizational structures are acting as a drag on that momentum. This disconnect is rarely harmless — it leads to frustration, frequent missed market opportunities, and strategic paralysis where leaders understand what the data indicates but lack the agility to respond. To close this gap, leaders must transform their organizations from rigid, slow-moving hierarchies into flexible, empowered networks able to move at the speed of the intelligence they use.

Problem 2 — Data Overload Without Understanding

Modern organizations are currently navigating an environment saturated with a relentless flow of information. Between real-time dashboards, advanced analytics platforms, AI-generated predictions, and automated recommendations, leaders are rarely lacking for data. However, there is a critical disconnect between the volume of information available and the depth of understanding achieved.

Research from the MIT Sloan Management Review highlights a significant friction point: **70% of executives** report that their organizations struggle to successfully translate this influx of data into actionable, meaningful decisions. In many cases, the abundance of information does not generate clarity; instead, it creates a state of "data paralysis." Leaders are frequently forced to navigate conflicting signals, rely on algorithmic outputs they do not fully grasp, and manage the intense pressure to act instantaneously, often without the necessary strategic context.

When leaders mistake mere data for genuine understanding, they shift from being proactive architects of their organization's future to being reactive responders to their own metrics. To reclaim the decision-making process, organizations must move beyond the accumulation of data and invest in the cognitive and structural capabilities required to synthesize that information into true intelligence.

Problem 3 — Speed Without Judgment

AI systems are engineered to optimize for quantitative metrics: peak efficiency, high prediction accuracy, and rapid decision-making. However, the core of effective leadership requires something fundamentally different—human judgment.

While algorithms excel at processing patterns, they lack the contextual nuance and moral compass required for complex, high-stakes decisions. Furthermore, research from Stanford's Human-Centered AI Institute demonstrates that these systems often inadvertently codify and scale the hidden biases present in historical data.

Without rigorous human oversight, organizations face cascading risks: institutionalized bias, sudden regulatory exposure, long-term reputational damage, and the erosion of public trust. Ultimately, the challenge we face is not a lack of technological capability; it is the strategic necessity of maintaining human accountability while machines accelerate the pace of decision-making. We must ensure that as we increase speed, we do not sacrifice the depth of human wisdom.

Problem 4 — The Emotional Shock of AI Transformation

When AI is introduced into an organization, it triggers more than just operational shifts; it fundamentally challenges professional identity. As workflows evolve, employees often grapple with existential questions about their value:

"Will AI render my expertise obsolete?"

"Does my role still have purpose?",

and "Where does the boundary between human judgment and algorithmic decision-making lie?"

The psychological weight of this transition is significant. According to PwC's *Future of Work* report, nearly **45% of employees** express anxiety regarding the impact of AI on their long-term career prospects.

When leaders overlook this emotional reality, adoption inevitably stalls. Resistance is rarely a critique of the software; it is a symptom of a lack of trust. Employees do not resist technology because they are inherently opposed to innovation—they resist it when they feel threatened, undervalued, or excluded from the transformation process. Bridging the gap between technical implementation and human confidence is not just a soft skill; it is a strategic imperative for successful AI integration.

Problem 5 — Organizations That Cannot Learn Fast Enough

AI is a landscape of perpetual motion. Models are refined, new tools emerge daily, and the functional capabilities of these systems expand at an unprecedented rate. Yet, in contrast to this technological agility, many organizations remain anchored to traditional leadership models built on the assumption of static expertise—the idea that a leader's value is derived from having all the answers based on experience.

This misalignment is costly. A comprehensive study on digital transformation from Harvard Business School revealed that over **70% of transformation initiatives fail**, largely because organizational leadership capabilities cannot keep pace with the speed of technical evolution.

In the AI era, relying on expertise alone is a liability. The shelf-life of knowledge is shrinking, turning "what I know" into a less relevant metric than "how quickly I can learn." To remain effective, leaders must transition from being holders of static expertise to becoming masters of continuous evolution. This requires a conscious, rigorous cycle of learning, unlearning outdated methodologies, and adapting strategies in real-time. In a world where the ground is constantly shifting, the most critical leadership skill is no longer command—it is adaptability.

Problem 6 — Culture Becomes the Hidden Barrier

Many AI initiatives begin as promising pilots, yet they struggle to scale. In fact, research from BCG indicates that only **26%** of companies successfully transition from experimental projects to achieving meaningful, organization-wide impact.

This friction rarely stems from the algorithms themselves; it is almost always a cultural challenge. While technology can be deployed rapidly, organizational behavior changes slowly. Companies often hit a wall because they fail to address deep-rooted structural issues, such as siloed teams, a lack of trust in automated systems, ambiguous decision ownership, and resistance to new ways of working. True scaling requires shifting the focus from technical deployment to cultural transformation.

Problem 7 — AI Without Human Partnership

Perhaps the most misunderstood challenge of AI is the relationship between humans and intelligent systems.

Many organizations treat AI in one of two ways:

- as a tool to automate human work
- or as a system whose outputs should be blindly trusted

Both approaches are flawed.

Research from **Accenture's Human + Machine report** shows that the highest-performing organizations are those that deliberately design **human-AI collaboration models**, combining machine efficiency with human creativity and judgment.

Without that balance, AI becomes either underutilized or dangerously over-trusted.

The Reality Leaders Are Beginning to See

Across industries, a pattern is emerging.

AI initiatives rarely fail because the technology does not work.

They fail because organizations have not yet developed the **human capabilities required to work intelligently with AI.**

Leadership assumptions built in the past are being quietly exposed.

Experience alone is no longer sufficient.

Authority alone no longer guarantees clarity.

Tool mastery no longer defines leadership capability.

The real transformation required in the AI era is not technological.

It is **human**.

Organizations that recognize this early are beginning to build new leadership capabilities that allow humans and intelligent systems to work together effectively.

Those who do not risk falling further behind as the pace of change continues to accelerate.

Why This Book Exists

According to BCG, the biggest challenge to GenAI ROI is not the technology—it's the people and the process. This type of change management is often overlooked, but top-performing organizations follow the 10-20-70 principle: **they dedicate 10% of their efforts to algorithms; 20% to data and technology; and 70% to people, processes, and cultural transformation.**

Deloitte's 2025 Global Human Capital Trends report shows that two-thirds of managers and executives say recent hires are not fully prepared — with "experience" being the most common shortfall — highlighting the need to focus on lasting human skills like curiosity and emotional intelligence.

Successful GenAI adoption depends on redesigning workflows, redefining roles, upskilling teams, and promoting collaboration between humans and AI. Without this change, even the most effective transformation efforts will fall short of their goals. Business leaders who invest in training and change management prepare their organizations to successfully accelerate generative AI implementation.

This book exists because the dominant narratives about AI leadership are incomplete.

Most books fall into one of two camps:

- **"AI will replace us"** — driven by fear, fatalism, and exaggeration
- **"AI is just another tool"** — driven by denial, oversimplification, and false comfort

Both miss the point.

AI is neither an existential enemy nor a neutral tool.

AI is a force multiplier.

It amplifies decision quality—or decision failure.

It accelerates learning—or locks in bad assumptions.

It strengthens cultures—or exposes their fractures.

The question is no longer *whether* AI will transform leadership.

The question is **whether leadership will evolve fast enough to guide it.**

This book is not about learning more AI.

It is about becoming the kind of leader—and building the kind of organization—that can **think, learn, decide, and adapt** alongside intelligent systems **without losing humanity, judgment, or trust.**

The Missing Layer: Integrated Intelligence

What AI exposes most brutally is that intelligence is not singular.

Organizations often over-index on one form of intelligence—technical, analytical, or operational—while neglecting others that now determine survival:

- **Sensemaking in ambiguity**
- **Emotional regulation under pressure**
- **Learning velocity over static expertise**
- **Ethical judgment at scale**
- **Cultural behavior change that actually sticks**
- **Human–AI collaboration across roles, geographies, and power structures**

AI-SENSE²: 7 Pillars

Pillar	Description
Sensemaking Intelligence	Navigate complexity, interpret information
Emotional & Social Intelligence	Regulate emotions, build trust
Neuro-Adaptive Learning	Accelerate skill acquisition, adapt quickly
Systemic Behavior & Culture	Engineer organizational norms, drive change
Ethical & Decision Intelligence	Make values-aligned decisions
Intercultural × Human-AI Intelligence	Lead across cultures, collaborate with AI
Foundation – Human + AI Synergy	Align human capabilities with AI collaboration

Without these, even the most advanced AI becomes noise.

This insight led to the creation of **AI-SENSE²**.

It is A Human-Centered Operating System for Intelligent Organizations

What readers will get?

Readers will understand why AI is not primarily a technology challenge but a **leadership bottleneck**. The introduction reframes the AI era as a test of human judgment, readiness, and maturity—setting the foundation for why leadership evolution, not tools, determines success.

PART I — The Leadership Shift AI Forces

Why yesterday's leadership model no longer works

What readers will get:

This opening part creates the necessary rupture before transformation can begin. It serves as the **unlearning phase**, deliberately destabilizing familiar leadership assumptions that no longer hold in an AI-driven world. Before leaders can build new capabilities through AI-SENSE², they must first let go of outdated beliefs about what makes leadership effective.

Here, the long-held myth that leadership equals experience, authority, or mastery of tools is dismantled. AI is introduced not as another technology to manage, but as a **cognitive force** that reshapes how thinking, decision-making, and power operate. As leaders confront this shift, a productive discomfort emerges—the realization that past success and accumulated expertise may no longer provide protection or relevance.

Although the framework has not yet been explicitly introduced, several **AI-SENSE²** pillars are activated beneath the surface. Leaders begin to unlearn fixed expertise, awakening **Neuro-Adaptive Learning**. They sense the growing mismatch between machine speed and human systems, engaging early **Sensemaking Intelligence**. They also recognize the hidden ethical risks of speed without judgment, hinting at the need for **Ethical & Decision Intelligence**.

The outcome of this part is readiness. Leaders become open, curious, and receptive to change. Their identity loosens, their certainty softens, and they become prepared to evolve. This psychological and cognitive reset makes everything that follows possible—and meaningful.

PART II — Sensemaking Intelligence

Making meaning in complexity before acting

What readers will get:

This part introduces the first explicit pillar of AI-SENSE^2: **Sensemaking Intelligence**. It establishes a foundational shift in leadership—from reacting to information toward intentionally creating meaning in complexity. In an environment saturated with data, dashboards, and AI-generated insights, leaders learn that understanding does not come from volume, but from interpretation.

Here, leaders develop awareness before action. Instead of rushing to optimize based on metrics, they learn to pause, question, and connect signals into coherent meaning. Data becomes a starting point for inquiry, not a substitute for judgment. Curiosity emerges as a critical leadership skill—the bridge that turns fragmented signals into insight and prevents premature conclusions.

Leaders also learn from intelligent systems themselves. Just as advanced AI systems sense patterns before acting, effective leaders must first understand context, relationships, and implications before making decisions. Seeing is no longer enough; meaning must be constructed.

This capability unfolds across layers. At the individual level, leaders examine their mental models and hidden assumptions. At the team level, they align narratives, so people are working from a shared understanding rather than conflicting interpretations. At the system level, they learn to navigate competing signals, informal power, and organizational politics without losing clarity.

The outcome is transformative. Leaders stop reacting to AI outputs as instructions and begin **framing reality correctly**, using intelligence—human and artificial—to guide thoughtful, context-aware action.

PART III — Ethical & Decision Intelligence

Judgment, responsibility, and when AI should pause

What readers will get:

As AI accelerates the pace and reach of decision-making, this part focuses on the most critical human responsibility that cannot be automated: **judgment**. Ethical & Decision Intelligence becomes the stabilizing force that ensures speed does not outpace wisdom.

Here, leaders develop deep **awareness of bias and moral clarity**, learning to recognize where data, models, or incentives may distort judgment. Decision-making is no longer about accepting the most accurate output, but about understanding context, consequences, and values. Human-in-the-loop accountability is deliberately designed so responsibility remains visible and shared, rather than hidden behind algorithms.

Governance evolves as well. Instead of rigid controls or reactive approvals, leaders learn to build **AI-augmented governance systems** that combine transparency, explainability, and ethical oversight with agility. Ethics becomes a daily leadership practice, not a compliance afterthought.

This capability unfolds across levels. At the individual level, leaders strengthen moral judgment and intuition. At the team level, accountability is distributed, encouraging collective ownership of ethical decisions. At the system level, governance frameworks are maturing to guide AI-driven decisions responsibly at scale.

The outcome is a decisive shift: leaders move beyond being approval bottlenecks and emerge as **ethical decision architects**—designing systems where intelligent speed and human wisdom advance together.

PART IV — Emotional & Social Intelligence

Regulating self, building trust, and sustaining energy

What readers will get:

Once leaders learn to see clearly, the next challenge is far more human: **stabilizing people in the presence of rapid AI-driven change**. This part builds the AI-SENSE^2 pillar of Emotional & Social Intelligence, recognizing that no technology scales successfully unless the human system can absorb it.

Here, leaders develop the ability to regulate their own emotional responses under uncertainty, pressure, and ambiguity. As AI reshapes roles, identities, and ways of working, emotions such as fear, resistance, and insecurity surface naturally. Leaders learn to address these responses directly—through presence, empathy, and trust—rather than bypassing them with logic or mandates.

Trust, psychological safety, and identity preservation become central leadership responsibilities. Teams adopt AI not because it is efficient, but because they feel safe enough to engage with it, question it, and learn alongside it. Adoption is revealed as an emotional process first, with rational acceptance following only after trust is established.

This capability unfolds across layers. At the self level, leaders regulate their nervous systems to remain calm and grounded. At the team level, they create psychological safety where concerns can be voiced without fear. At the system level, culture evolves to support people through transformation rather than overwhelm them.

The outcome is decisive: **AI stops being feared or resisted**. Leaders become emotionally trustworthy anchors—steady, credible, and capable of guiding humans through change before scaling machines.

PART V — Neuro-Adaptive Learning

Learning faster than change

What readers will get:

This part redefines what it means to be a capable leader in the AI era. As AI evolves at an unprecedented pace, expertise alone is no longer enough. Leaders must develop the ability to **outlearn the rate of change itself**. Neuro-Adaptive Learning becomes the core capability that keeps leadership relevant when knowledge expires quickly.

Here, the emphasis shifts from what leaders know to **how they learn**. Learning velocity matters more than accumulated expertise. Leaders practice presence instead of reaction, creating space for reflection even as algorithms move at speed. They learn to enter cognitive flow under acceleration—thinking deeply without being overwhelmed—and to build continuous learning loops that turn every outcome into feedback.

This transformation unfolds across levels. At the individual level, leaders develop cognitive flexibility, letting go of rigid mental models and embracing unlearning as a strength. At the team level, learning becomes collective through shared experimentation and fast feedback. At the system level, knowledge is transferred and scaled so learning does not remain trapped in silos.

The result is a fundamental identity shift—from **"I know"** to **"I can adapt."** Leaders remain effective not by keeping up with AI, but by evolving alongside it, month after month.

PART VI — Systemic Behavior & Culture

Embedding change so it sticks

What readers will get:

This part is where AI-SENSE² shifts from individual leadership capability to **organizational reality**. It addresses the hardest truth of transformation: AI succeeds or fails not because of algorithms, but because of the behaviors and culture that surround them.

Here, the focus moves to **systemic behavior**—the everyday habits, rituals, and norms that quietly determine whether AI is adopted, trusted, and sustained. Leaders learn to shape identity-aligned habits at the individual level, reinforce shared norms within teams, and deliberately engineer culture at the system level. Transformation is approached behavior-first, ensuring that new ways of working take root before technology is scaled.

Coaching emerges as a critical infrastructure rather than a support function. It becomes the mechanism through which behaviors are practiced, reinforced, and corrected over time. Instead of relying on mandates or one-time change programs, leaders create feedback loops that allow AI-enabled ways of working to become natural and repeatable.

The core insight is unmistakable: **AI doesn't fail technically—it fails systemically**. When culture is ignored, even the most advanced solutions remain isolated experiments. When culture is designed intentionally, AI integrates seamlessly into daily work.

The outcome is profound and practical. AI stops being a side project or innovation initiative and becomes **"the way we work around here"**—embedded into decisions, collaboration, and execution at scale.

PART VII — Human-AI Synergy

Working with AI without losing what makes us human.

What readers will get:

This part marks a fundamental shift in how leaders relate to artificial intelligence. Instead of viewing AI as something to control, automate, or defend against, leaders learn to build a **true partnership** with intelligent systems. This is where the AI-SENSE² pillar of **Human-AI Synergy** fully takes shape.

Here, collaboration replaces control. Leaders develop the ability to delegate, cognitively deciding what thinking should be augmented by AI and what must remain human. Workflows are redesigned so humans and machines contribute where each is strongest, creating flow rather than friction. Trust between people and intelligent systems is deliberately built, not assumed, allowing AI insights to be questioned, refined, and responsibly acted upon.

At the same time, leadership expands to include both **cultural and algorithmic dimensions**. Leaders learn to guide not only human behavior and norms, but also how algorithms influence decisions, incentives, and outcomes. Human judgment remains central, but it is now amplified by machine intelligence.

The outcome is clear and practical: AI stops being a threat to identity or a crutch for thinking. It becomes a **co-pilot**—enhancing clarity, accelerating insight, and enabling leaders to operate at a higher level without losing what makes leadership human.

PART VIII — Intelligence at Enterprise Scale

From pilots to platforms, from experiments to advantage

What readers will get:

This part demonstrates that AI-SENSE² is not a conceptual model or a leadership philosophy limited to individuals or small teams. It shows how the framework operates at **enterprise scale**, across complex, global systems where coordination, culture, and consistency matter as much as technology.

Here, the focus shifts to **system-level orchestration**. AI-SENSE^2 is applied across geographies, functions, and operating models, proving that intelligent leadership is not dependent on proximity or uniformity. When culture and leadership maturity are intentionally developed, intelligence scales naturally—without creating fragmentation or rigidity.

What emerges is a clear pattern: organizations that treat AI as a series of pilots remain stuck in experimentation, while those that apply AI-SENSE^2 turn isolated successes into repeatable platforms. Leadership alignment and cultural coherence become the multipliers that convert innovation into sustained advantage.

The outcome is decisive. AI-SENSE^2 moves beyond theory and experimentation to become a **practical enterprise capability**, enabling organizations to scale intelligence confidently and compete effectively in a globally connected, AI-driven world.

PART IX — The AI-SENSE^2 Leader

Integration, future readiness, and leadership evolution

What readers will get:

This final stage represents the full integration of AI-SENSE^2 into a new leadership identity. At this point, the framework is no longer something leaders *use*—it becomes how they **think, decide, and lead**. All seven pillars converge into a single, coherent leadership operating system that guides behavior naturally, not mechanically.

The AI-SENSE^2 Leader learns to **sense before acting**, reading complexity with clarity instead of reacting to noise. They regulate their inner state before influencing others, creating emotional stability in fast-changing environments. Learning becomes continuous, not episodic; curiosity replaces certainty, and adaptation becomes instinctive. Culture is no longer treated as a soft concept but as a living system shaped through everyday habits, rituals, and decisions. Ethics is no longer a policy layer—it becomes an internal compass guiding judgment under speed and pressure. AI is no longer a tool to control or fear, but a partner to think with, enabling better decisions and deeper insight. Intelligence is no longer local—it scales across teams, systems, and geographies as a shared capability.

The final transformation is not just personal growth, but systemic evolution: **individuals become intelligent leaders, teams become intelligent collectives, and organizations become adaptive, resilient systems**—ready not just for today's AI world, but for whatever comes next.

Old Leadership vs. AI-Age Leadership

Characteristic	Old Leadership	AI-Age Leadership
Knowledge	I know	I learn
Approach	control	adapt
Mindset	certainty	curiosity
Skill	expertise	learning velocity

The Leaders Who Will Matter Most:

What readers will get:

Readers will leave with a clear understanding of what defines enduring leadership in the AI era: the ability to **sense, learn, decide, and lead with AI—without losing humanity**. The conclusion reinforces that the future belongs not to the fastest adopters, but to the most grounded, adaptive, and human-centred leaders.

What Is AI-SENSE²?

AI-SENSE² is not a model.

It is not a framework layered on top of existing leadership theory.

It is a leadership operating system for an AI-driven world.

AI-SENSE² integrates **human intelligence, neuroscience, organizational behavior, and AI collaboration** into a coherent system that works across three levels:

- **Self** — how leaders think, feel, decide, and learn
- **Team** — how people collaborate, trust, and adapt together
- **System** — how culture, incentives, governance, and scale reinforce behavior

The framework is built on **seven integrated pillars**:

1. **Sensemaking Intelligence** — making meaning before acting
2. **Emotional & Social Intelligence** — regulating self and building trust
3. **Neuro-Adaptive Learning** — learning faster than change
4. **Systemic Behavior & Culture** — embedding habits that stick
5. **Ethical & Decision Intelligence** — knowing when speed must pause
6. **Intercultural & Inclusion Intelligence** — leading across differences
7. **Human–AI Synergy** — partnering with machines without surrendering judgment

The "2" matters.

It represents the expanded intelligence required in a connected, AI-augmented world—where leaders must operate across cultures **and** with intelligent systems simultaneously.

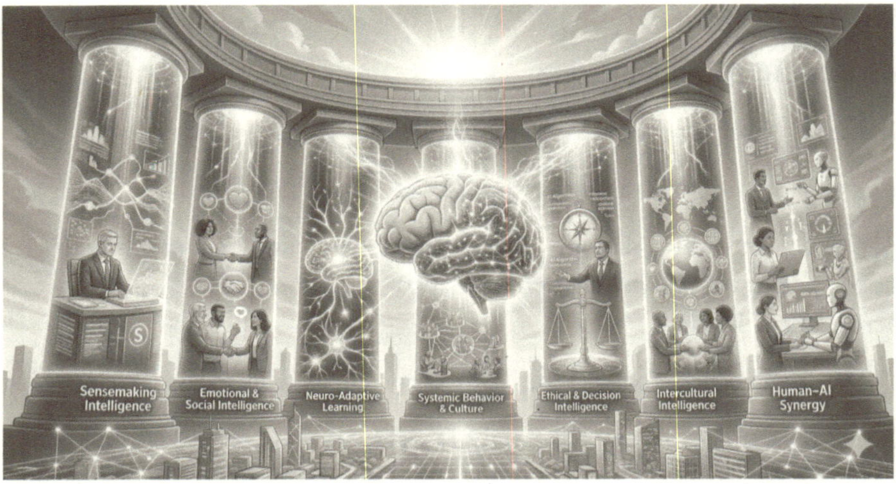

AI-SENSE^2 is not a theory. It Is Built on Evidence.

AI-SENSE^2 was not created in isolation.

It integrates decades of research across leadership science, neuroscience, organizational behavior, and AI transformation studies.

AI Success Is 70% Human, 30% Technical

BCG – The 10–20–70 Rule

Boston Consulting Group's research on AI transformations shows:

- 10% of effort should focus on algorithms
- 20% on data and technology
- 70% on people, processes, and culture

Organizations that overinvest in technical build and underinvest in leadership and operating model redesign consistently fail to scale AI.

AI-SENSE^2 focuses explicitly on that 70%.

It strengthens:

- Leadership sensemaking
- Governance discipline

- Learning velocity
- Cultural alignment
- Ethical oversight
- Human–AI workflow integration

This is not a philosophical positioning. It aligns directly with documented transformation outcomes.

AI Maturity Research: What Differentiates Leaders

McKinsey Global AI Survey Findings

Top AI-performing organizations are more likely to:

- Embed AI into core workflows (not side experiments)
- Establish formal AI governance structures
- Monitor model performance continuously
- Invest in leadership capability development
- Align AI strategy with business value metrics

These success drivers map directly to AI-SENSE^2 pillars:

- Systemic Behavior & Culture
- Ethical & Decision Intelligence
- Neuro-Adaptive Learning
- Human–AI Synergy

The evidence shows: AI maturity is systemic maturity.

MIT Sloan & BCG: The Cultural Factor

Research from MIT Sloan and BCG demonstrates:

High AI-performing organizations exhibit:

- Strong cross-functional collaboration
- Executive-level AI fluency

- Psychological safety for experimentation
- Structured decision governance
- Organizational learning loops

These elements are not technical architecture components.

They are leadership and cultural capabilities.

AI-SENSE² integrates these capabilities into one operating system.

Why AI Projects Fail: Evidence-Based Patterns

Across industries, AI projects fail for predictable reasons:

- No clear data ownership
- Weak governance frameworks
- Lack of monitoring and drift management
- Poor cross-team alignment
- Executive misunderstanding of trade-offs
- Cultural resistance to AI adoption
- Ethical oversight was added too late

Notice what is missing from that list:

Algorithmic failure.

Most AI failures are systemic, not mathematical.

AI-SENSE² directly addresses those systemic failure modes.

Responsible AI & Governance Research

Global frameworks, including:

- OECD AI Principles
- EU AI regulatory guidance
- NIST AI Risk Management Framework

All emphasize:

- Human accountability
- Transparency
- Bias monitoring
- Ongoing risk assessment
- Lifecycle governance

AI-SENSE² embeds Ethical & Decision Intelligence into leadership practice, ensuring governance is built in—not layered after deployment.

Neuroscience & Learning Evidence

Modern neuroscience confirms:

- Cognitive flexibility predicts adaptive performance
- Emotional regulation improves decision quality under pressure
- Psychological safety increases innovation velocity
- Continuous learning loops outperform static expertise

AI-SENSE²'s Neuro-Adaptive Learning and Emotional & Social Intelligence pillars are grounded in these principles.

In fast-moving AI environments, learning velocity matters more than accumulated expertise.

Research supports this shift.

Human–AI Collaboration Research

MIT Center for Collective Intelligence and multiple enterprise AI studies show:

The highest performance gains occur when:

- Humans and AI divide cognitive labor clearly
- AI augments decision-making rather than replaces it

- Leaders understand AI limitations
- Workflow redesign accompanies model deployment

AI-SENSE² explicitly addresses Human–AI Synergy at self, team, and system levels.

This is not about building AI faster.

It is about collaborating with AI intelligently.

Integrated Intelligence: The Gap AI-SENSE² Closes

Most organizations treat these domains separately:

- AI governance in one function
- Leadership development in another
- Data science in another
- Culture change as a side initiative

AI-SENSE² integrates them.

It unifies:

Human Leadership

Neuroscience

Organizational Behavior

AI Governance

Enterprise Execution

This integration is what enables scalable AI maturity.

What AI-SENSE² Does (And Does Not Claim)

AI-SENSE² does not claim to build better neural networks.

It claims—and evidence supports—that:

AI initiatives succeed when leadership coherence increases alongside technical intelligence.

When intelligence rises without coherence, failure accelerates.

When intelligence rises with coherence, transformation stabilizes.

This is the central thesis supported by research across consulting firms, academic institutions, and global AI governance bodies.

AI-SENSE² stands on four evidence pillars:

1. AI transformation research confirms human & system factors drive ROI.
2. Organizational science shows culture and governance determine scalability.
3. Neuroscience proves that adaptive learning and emotional regulation improve performance under uncertainty.
4. Responsible AI frameworks globally require integrated leadership oversight.

AI-SENSE² integrates all four into one leadership operating system.

That is its foundation.

That is its credibility.

That is its evidence.

How This Book Is Structured

This book is deliberately designed to show how genuine transformation happens. Change doesn't occur through a single announcement, a top-down order, or a new tool introduction. It starts inside the leader and spreads outward. It begins with mindset and skills, then influences behavior, then alters teams, and finally changes systems.

As you move through the nine parts of the book, the journey mirrors that progression. You begin by examining why traditional leadership models begin to crack under the pressure of AI. From there, you learn how to make sense of complexity before taking action, especially in environments flooded with data and speed. The focus then deepens into judgment, ethics, and emotional regulation—because decisions made under pressure shape trust and outcomes more than algorithms do.

Next, the book examines how continuous learning, reinforced behaviors, and intentional culture design enable transformation to expand beyond individuals. It then explains how a human–AI partnership can become a true competitive advantage rather than a source of tension. Finally, the journey ends with understanding how intelligence shifts from isolated pilots to enterprise-wide impact.

Each chapter features real leadership stories, research-based insights, practical practices, and clear connections to the AI-SENSE^2 capabilities. This is not just abstract theory. It is rooted in what actually causes failure—and what truly succeeds—inside organizations navigating AI transformation in real time.

Who This Book Is For?

This book is written for leaders who sense that something fundamental has shifted, even if they can't yet clearly express what it is. They feel it in the speed of decisions, the weight of responsibility, and the widening gap between how leadership used to operate and what is now being asked of them.

It speaks to senior leaders navigating AI-driven transformation without a clear map, to product, technology, and transformation leaders who feel stretched across competing demands, and to coaches and consultants who are supporting leaders through constant uncertainty. It is also for boards that are beginning to ask more difficult questions about risk, ethics, and long-term value, and for high-potential leaders who are preparing for a future that shows no sign of slowing down.

This isn't a book full of step-by-step AI implementation guides. It doesn't include tool comparisons, vendor frameworks, or promises of easy automation. It also doesn't suggest that complexity can be removed with a simple checklist.

Instead, this book is for those seeking clarity in the midst of acceleration, confidence without the illusion of certainty, and a form of leadership that scales with intelligence rather than pushing against it. If that is the kind of leadership you are looking to grow into, then you are in the right place.

A Different Kind of Leadership Challenge:

This is not a book about tools. It is not a guide to prompts, platforms, or predictions. It does not try to keep pace with technologies that will change again before the ink dries. Instead, it focuses on something far more enduring: *who you become as a leader when machines begin to think faster than you ever could.*

The book explores what it means to lead when certainty disappears, to make decisions when speed itself can become dangerous, and to remain deeply human in a world where intelligence is accelerating at an unprecedented rate. It is about evolving your leadership in a way that expands your capacity without requiring you to lose yourself in the process.

The moment AI made my old way of leading obsolete did not mark the end of my relevance. It marked the beginning of a deeper responsibility. This book is an invitation to step into that responsibility and to walk a journey of leadership that is wiser, more grounded, and fit for the future we are already living in.

A Final Orientation Before You Begin

AI will continue to evolve—faster than any leadership training cycle, governance model, or organizational habit we currently rely on. That pace is not something leaders can slow down or control. What *is* within reach is far more important.

Leaders can choose how they interpret the signals emerging from complexity, how they regulate fear and pressure when decisions feel compressed, and how they recognize the moments when speed must pause in service of wisdom. They can decide how intentionally they learn, unlearn, and relearn, how their everyday behavior shapes culture, and how they partner with machines without surrendering judgment.

This book is an invitation to do that work deliberately. Not reactively. Not defensively. But with clarity and responsibility.

Because the future will not belong to the leaders who know the most about AI. It will belong to those who can think clearly amid acceleration, learn continuously without ego, decide wisely under pressure, and lead human beings in a world where intelligence is everywhere—but wisdom remains rare.

That journey does not begin someday.

It begins now.

How I Developed the AI-SENSE² Framework

The AI-SENSE² framework did not emerge from theory alone. It is the result of years spent building, applying, and stress-testing leadership, coaching, innovation, and agility frameworks across real people, real teams, and real organizational constraints.

As **Chandan Lal Patary**, my work has always sat at the intersection of **human potential and systemic complexity**. Over time, I developed multiple frameworks—IMPACT, SPARK, CREATE, 7E Leadership, and coaching models for individuals, teams, and enterprises. Each framework solved a specific leadership or transformation problem. Yet as AI entered the workplace at scale, I began to notice a deeper pattern.

Leaders were not struggling because they lacked tools. They were struggling because their **thinking models were no longer sufficient**.

AI exposed something fundamental: leadership frameworks that worked in stable or slowly changing environments began to fracture under speed, uncertainty, and machine intelligence. Traditional agility focused on execution. Innovation frameworks focused on ideas. Coaching frameworks focused on growth. But none fully addressed what happens when **machines think faster than humans—and humans must still decide**.

AI-SENSE² emerged as an integration, not an invention.

It combines sensemaking, ethical judgment, emotional regulation, learning agility, and systemic leadership into a single operating system for the AI era. Every element of the framework is grounded in lived experience—coaching leaders through uncertainty, guiding teams through overload, and helping organizations move from AI experimentation to real impact.

Rather than asking, *"How do we use AI better?"*, AI-SENSE² asks a more human question:

"How must leadership evolve so intelligence—human and artificial—creates meaning, trust, and progress?"

This framework is not about replacing leadership with technology. It is about **elevating leadership maturity,** so AI becomes a partner, not a crutch.

AI-SENSE² embodies everything I have learned over decades of developing frameworks—refined through coaching, tested in complex situations, and driven by one core belief:

The future of leadership does not belong to those who think fastest,

but to those who can sense deeply, decide wisely, and lead humanely—

even when machines think faster.

The AI-SENSE² Framework: Leading Humans, Culture, and Systems in the Age of AI

The AI-SENSE² Framework

Leading Humans, Culture, and Systems in the Age of AI

Every major shift in leadership history has required a new way of thinking.

The Industrial Age demanded operational discipline.

The Information Age rewarded analytical intelligence.

The Digital Age elevated speed, execution, and scale.

The AI Age changes the equation entirely.

For the first time, leaders are not the fastest processors in the system.

They are not the primary source of analysis.

They are not even the most "intelligent" entity in the room—at least not in the traditional sense.

And yet, leadership has never mattered more.

Because when intelligence becomes abundant, **judgment becomes scarce**.

When automation accelerates, **meaning becomes fragile**.

When systems think faster, **humans must think deeper**.

This is the context in which **AI-SENSE²** was created.

Why a New Leadership Framework Is Necessary

For many years, most leadership frameworks were designed for a very different world. In that world, change happened slowly. Organizations had time to adjust, learn, and stabilize before the next wave of change arrived. Expertise also grew gradually. People could spend decades mastering a field, and the knowledge they gained remained valuable for a long time. Decisions moved through organizations step by step, scaling in a predictable and linear way. Culture evolved gently as people adapted to new circumstances over time. Technology typically followed strategy; leaders decided what direction to take, and technology was implemented afterward to support that plan.

Artificial intelligence changes all of these assumptions.

In environments shaped by AI, change is no longer occasional—it is constant and accelerating. New tools, capabilities, and ways of working appear continuously, forcing organizations to adapt far more quickly than before. Expertise no longer lasts for decades; knowledge can become outdated in a short period as technologies evolve, and new methods emerge. Decisions that once spread slowly through layers of management can now scale instantly through algorithms, automated systems, and data platforms. Organizational culture is no longer something that shifts gradually. Instead, it is tested every day as teams face rapid change, uncertainty, and new forms of collaboration between humans and machines. In this new reality, technology no longer simply supports strategy—it actively reshapes it in real time.

What often fails in this environment is not intelligence or talent. Leaders are capable and experienced. The deeper challenge lies in integration. The capabilities needed to lead in the age of AI are scattered across different fields of knowledge. Insights from neuroscience may explain how humans learn and adapt. Leadership theory offers models for guiding people through change. AI ethics raises critical questions about responsibility and trust. Organizational behavior explores how teams function and how culture evolves. Yet these areas are often studied and applied separately.

As a result, leaders are left trying to piece together fragments of knowledge that were never designed to work as a unified system. The real challenge of leadership in the AI era is not simply learning new tools but integrating these diverse capabilities into a coherent way of thinking and acting.

AI-SENSE^2 integrates what has been artificially separated.

What AI-SENSE^2 Actually Is

AI-SENSE^2 is a leadership operating system for an AI-driven world.

It integrates:

- **Human leadership** (judgment, trust, meaning, ethics)
- **Neuroscience** (how humans actually think, feel, learn, and adapt)
- **AI collaboration** (working *with* intelligent systems, not around them)

- **Enterprise-scale impact** (culture, systems, governance, and execution)

Its purpose is not to make leaders more technical.

Its purpose is to make leaders **more adaptive, more grounded, and more capable of orchestrating intelligence at scale**—human and artificial.

The Leadership Questions AI-SENSE² Was Built to Answer

Every meaningful leadership framework begins with questions—not answers.

AI-SENSE² exists because leaders across industries are confronting the same silent tensions, often without the language to name them. These are not technical problems. They are **human, systemic, and ethical challenges** created by intelligence that now moves faster than organizations were designed to adapt.

Below are the **most powerful questions and challenges** AI-SENSE² addresses—explicitly, systematically, and at scale.

1. Sensemaking Challenges

As AI systems generate more dashboards, forecasts, and recommendations, leaders often face a surprising paradox: the more data they receive, the harder it becomes to truly understand what is happening. Signals conflict. Metrics point in different directions. One model suggests growth while another warns of risk. Instead of clarity, there is noise.

This raises difficult questions. How do you make sense of reality when the information is overwhelming or inconsistent? How do you distinguish between random fluctuations and early warning signs? When dashboards disagree, who decides what actually matters? And how do leaders ensure that teams across functions, cultures, and power structures interpret the same data in a shared way?

The core challenge is this: AI can generate answers, but it cannot provide context on its own. It does not understand organizational history, political dynamics, or human consequences. Leaders must decide what those answers mean before acting on them.

AI-SENSE² strengthens this capability. It teaches leaders to interpret ambiguity before accelerating action. It helps them align narratives across teams so that data leads to shared understanding rather than fragmented reactions. And it encourages slow thinking just enough to avoid decisions made at dangerous speed. In a world saturated with information, the real advantage is not more data, but clearer meaning.

2. Emotional & Social Challenges

As AI reshapes roles, processes, and expectations, the pace of change often rises faster than people can comfortably handle. Deadlines compress. Decisions multiply. Job identities feel uncertain. Even confident teams can begin to feel stretched. When everything feels urgent, emotional capacity becomes the real constraint.

Leaders face difficult questions in these moments. How do you stay calm when pressure is constant? How do you prevent fear, resistance, or quiet burnout from undermining AI adoption? How do you build trust when people worry about being replaced? And how do you guide others through uncertainty without pretending to have complete certainty yourself?

The challenge is deeply human. AI can transform work in months, but the human nervous system cannot recalibrate at the same speed. Without emotional regulation, urgency spreads. Anxiety becomes contagious. Trust weakens.

AI-SENSE² empowers leaders in this field. It aids them in managing their emotional responses during prolonged uncertainty, preventing stress from passing through the system. It fosters psychological safety, where questions and concerns can be shared openly. Additionally, it helps leaders turn anxiety into curiosity, transforming uncertainty into engagement rather than resistance. In this way, AI adoption becomes not only a technical change but a human transition guided by steadiness and care.

3. Learning & Adaptation Challenges

In today's AI-driven world, expertise does not last as long as it once did. Skills that took years to build can become outdated in months. This creates a quiet anxiety for many leaders: how do you stay relevant when what made you successful is no longer enough?

The real challenge is not learning something new once in a while. It is learning continuously without slowing down execution. It is having the courage to let go of methods, assumptions, and habits that once worked but now hold you back. Organizations, too, must move beyond simply collecting knowledge. They must learn how to scale learning itself — how to adapt as a system, not just as individuals.

AI-SENSE² addresses this challenge by shifting the focus from mastery to adaptability. It builds neuro-adaptive learning habits that strengthen cognitive flexibility. Instead of clinging to expertise, leaders develop learning velocity — the ability to evolve quickly, reflect deeply, and adjust without burning out. In an environment where change is constant, the true advantage is not what you know, but how fast and wisely you can learn.

4. Systemic & Cultural Challenges

Even the best AI strategy will fail if behavior does not change with it. Many organizations launch successful AI pilots, celebrate early wins, and then watch momentum fade. People quietly return to old habits, especially under pressure. Incentives reward familiar behaviors. Meetings run the same way. Decisions follow the same patterns. The strategy changes, but the culture does not.

This is the hidden truth of transformation: technology moves fast, but culture moves slower — and culture ultimately decides the outcome. If systems, rewards, and everyday routines remain unchanged, AI becomes an add-on instead of a new way of working.

AI-SENSE² addresses this gap by focusing on behavior, not just vision. It helps leaders design systems that reinforce the right actions. Incentives, rituals, workflows, and feedback loops are aligned so that new behaviors feel natural and supported. Instead of relying on slogans or one-time change programs, AI adoption becomes embedded in how the organization operates. Over time, what once felt like an initiative becomes part of the culture itself.

5. Ethical & Decision Challenges

As AI accelerates decisions, a new tension emerges: what happens when speed outpaces judgment? Algorithms can analyze patterns and generate recommendations in seconds, but they do not understand consequences the way humans do. They do not feel the weight of impact. They do not carry responsibility.

This raises deeper leadership questions. When should AI be allowed to proceed, and when must it pause? Who is accountable if an AI-informed decision harms a customer, employee, or community? How do we govern choices that scale instantly across thousands or millions of people? And how do we balance efficiency with dignity, trust, and fairness?

The core challenge is clear. AI can optimize processes, reduce costs, and increase accuracy. But it cannot take moral responsibility. That remains human.

AI-SENSE² empowers leaders in this field. It develops ethical reasoning amid uncertainty, helping them pause just enough to make smart choices. It creates human-in-the-loop decision frameworks, ensuring accountability is never delegated solely to machines. Plus, it promotes responsibility at every level — individual, team, and system — so that as intelligence advances, judgment advances with it.

6. Human–AI Collaboration Challenges

As machines begin to think faster than people, leadership enters unfamiliar territory. AI can process vast amounts of information in seconds, generate options instantly, and surface insights that would take humans days or weeks to uncover. But speed alone does not define good leadership. The real question becomes: what should humans still decide, and what should AI handle?

There is also a hidden risk. When intelligent systems perform more of the cognitive work, people may begin to rely on them too heavily. Over time, critical thinking can weaken. Judgment can dull. Trust in algorithms can quietly replace thoughtful evaluation. Leaders must learn how to collaborate with AI without becoming dependent on it.

The challenge is not to compete with machines, but to partner with them wisely. Leaders must redesign workflows so that humans and AI each contribute where they are strongest. They must delegate cognitive tasks without delegating responsibility. And they must protect human wisdom even as they accelerate insight. **AI-SENSE²** guides leaders through this balance, ensuring that speed enhances judgment rather than replacing it.

7. Intercultural & Global Challenges

When intelligence begins to scale across borders, leadership becomes more complex than ever. AI systems trained in one context may behave differently in another. Cultural norms, language nuances, regulatory expectations, and social values can all influence how technology is interpreted and applied. What works seamlessly in one country may create friction or misunderstanding in another.

This raises important questions. How do we ensure AI does not unintentionally amplify bias on a global scale? How do teams across regions collaborate effectively when they are relying on the same intelligent systems? And how do leaders design digital operating models that respect diversity rather than flatten it?

The challenge is subtle but powerful. AI can amplify cultural assumptions at machine speed. If bias exists in data, design, or decision logic, it can spread instantly and globally.

AI-SENSE² addresses this by strengthening intercultural leadership at scale. It helps leaders design inclusion into systems from the beginning, rather than correcting issues after harm occurs. It builds trust across geographies, ensuring that intelligence can scale without eroding respect, fairness, or cultural understanding.

8. Leadership Identity Challenges

There was a time when leadership meant having the answers. Experience, expertise, and decisiveness defined authority. But in an AI-driven world, machines often process more data, surface more insights, and analyze more variables than any individual leader can. This shifts the ground beneath traditional leadership identity.

Leaders are now faced with uncomfortable questions. What does it mean to lead when AI may know more about patterns, predictions, or probabilities than you do? How do you maintain credibility without pretending certainty? How do you guide people through complexity not by delivering answers, but by asking better questions? And how do you remain deeply human in systems increasingly shaped by machines?

The old model—expert, controller, ultimate decider—no longer fits. Clinging to it creates rigidity and fear.

AI-SENSE² helps leaders redefine authority. Authority becomes the ability to make sense of complexity, apply judgment where it matters, and create coherence in uncertainty. Leadership shifts from being the source of answers to being the architect of learning. Presence replaces performance. Curiosity replaces defensiveness. Instead of acting as answer machines, leaders become learning architects—designing environments where people and AI think better together.

The Unifying Challenge AI-SENSE² Addresses

At its core, AI-SENSE² answers one defining question:

How do we scale intelligence without losing humanity?

Or more practically:

How do leaders think, decide, learn, and lead when intelligence is abundant—but wisdom is not?

Why These Questions Matter

Because organizations don't fail in the AI age due to:

- lack of data
- lack of models
- lack of technology

They fail because:

- humans are overwhelmed
- systems are misaligned
- leaders are unprepared for the cognitive, emotional, and ethical load AI creates

AI-SENSE² exists to close that gap—**deliberately, systematically, and humanely**.

AI-SENSE²: Pillars of AI Agility

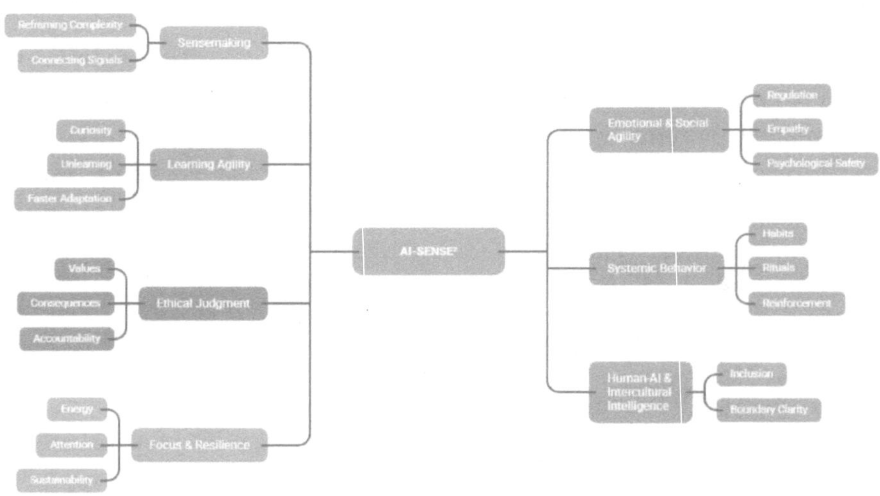

The Core Design Principle: Intelligence Must Operate at Three Levels

AI-SENSE² is built on a foundational insight confirmed by decades of organizational research:

Sustainable transformation only happens when leadership capability develops simultaneously at **three levels**.

1. Self

How leaders think, feel, decide, and learn under pressure.

2. Team

How people collaborate, trust, challenge, and adapt together.

3. System

How culture, incentives, governance, and structures reinforce—or sabotage—intended behavior.

Most leadership programs overinvest in the **Self**, underinvest in the **Team**, and almost ignore the **System**.

AI-SENSE² explicitly connects all three.

The Meaning of SENSE²

The name **AI-SENSE²** is intentional.

SENSE

Represents the core human intelligence leaders must develop to remain effective:

- **Sensemaking**
- **Emotion**
- **Neuro-adaptive learning**
- **Systemic behavior**
- **Ethical decision-making**

These capabilities govern how humans perceive reality, regulate themselves, and act wisely under complexity.

² (Squared)

Represents the *expanded intelligence* required in a connected world:

- **Intercultural Intelligence**
- **Human–AI Intelligence**

The square signifies multiplication, not addition.

When these dimensions are integrated, leadership capacity compounds rather than fragments.

The Foundation: Human + AI Synergy

Before exploring the pillars, one foundation must be clear:

AI-SENSE² does not treat AI as a tool.

It treats AI as a cognitive partner.

This does not mean surrendering judgment to machines.

It means understanding:

- what AI does exceptionally well
- where AI is blind
- where human judgment must remain absolute

Why This Foundation Matters

Organizations rarely fail with AI because the models are weak. More often, they fail because the surrounding system is not ready. Workflows remain unchanged, decision rights are unclear, ethical considerations are added too late, and humans are either excluded from the loop or silently overridden by automated outcomes.

AI-SENSE² begins by addressing this gap. It focuses on establishing a responsible partnership between humans and machines. Leaders develop a realistic understanding of what AI can and cannot do. Teams redesign their workflows so that intelligence is augmented rather than automated blindly. Organizations put governance and ethical guardrails in place early, not as an afterthought. Most importantly, AI is positioned as a decision-support system, while humans remain clearly accountable for the consequences of those decisions.

This foundation ensures that intelligence does not replace judgment but strengthens it—allowing AI to amplify human wisdom rather than undermine it.

The Seven Pillars of AI-SENSE²

Each pillar operates across **Self** → **Team** → **System**, ensuring individual growth translates into organizational impact.

AI-SENSE² Framework

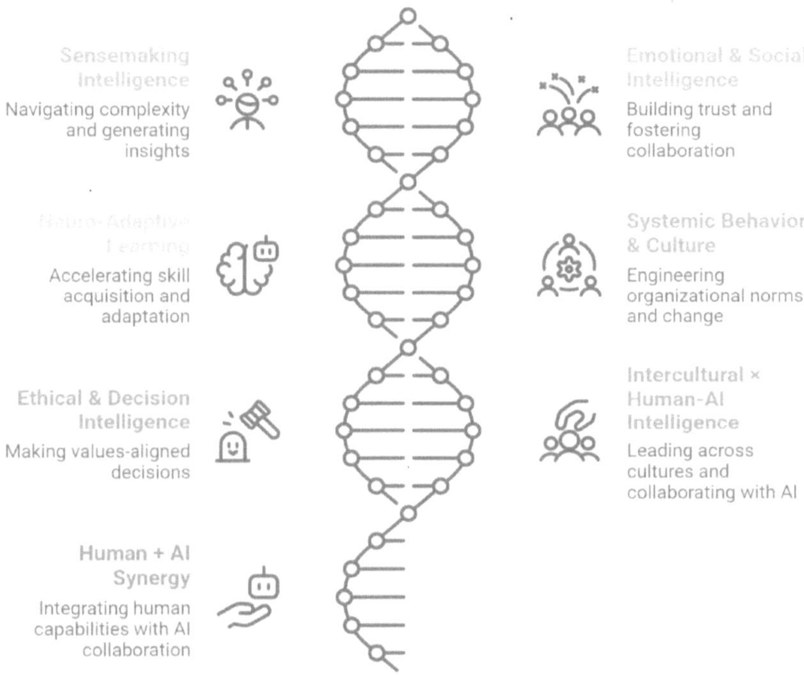

1. Sensemaking Intelligence

Making Meaning in Complexity

AI fills organizations with more data than ever before. Dashboards update in real time. Models generate predictions instantly. Signals arrive from every direction. But more data does not automatically lead to better understanding.

In fact, without sensemaking, data often creates confusion.

This is where leadership matters most. Leaders must slow down before acting. They must interpret ambiguity, notice contradictions, and ask what the data actually *means* in context. Sensemaking helps leaders turn scattered information into a shared understanding that people can align around.

At the individual level, this begins with self-awareness. Leaders examine their assumptions, question their mental models, and recognize blind spots that shape how they interpret information. They learn to separate signal from noise.

At the team level, sensemaking brings people into the same story. Different perspectives are surfaced, competing interpretations are discussed, and a shared narrative begins to form. This alignment reduces friction and prevents teams from pulling in different directions.

At the system level, sensemaking helps leaders navigate informal power structures, hidden politics, and unspoken decision paths. It reveals how decisions actually get made—not just how they are supposed to be made.

Without sensemaking, speed becomes risky.

With sensemaking, clarity emerges—and action becomes intelligent rather than reactive.

2. Emotional & Social Intelligence

Regulating Self, Reading Others, Building Trust

As AI enters everyday work, leaders are asked to think faster, decide under pressure, and absorb more information than ever before. This increase in speed and complexity also increases emotional strain. When leaders are not emotionally regulated, stress spreads quickly. Fear rises, trust drops, and teams begin to resist change—not because they don't understand AI, but because they don't feel safe.

This is why emotional and social intelligence matter so deeply in the AI age.

When leaders learn to regulate themselves, they stay calm under pressure and manage stress instead of passing it on. Their steadiness creates space for others to think clearly. Teams feel safe to speak up, question AI outputs, admit uncertainty, and learn together. Trust begins to scale—not through rules, but through consistent emotional presence.

At the individual level, this capability strengthens emotional resilience and self-awareness. At the team level, it builds psychological safety, empathy, and healthy emotional influence. At the system level, it aligns culture during transformation, preventing fear from turning into silent resistance.

Ignored emotion doesn't disappear.

It hardens into resistance.

But when emotion is understood and managed well, it becomes momentum—fuel for learning, trust, and lasting change.

3. Neuro-Adaptive Learning

Learning Faster Than Change

In AI-driven environments, the pace of change is relentless. Skills lose relevance faster than careers were ever designed to adapt. What worked last year may already be outdated. In this reality, experience alone is no longer enough to stay effective.

This is why learning speed has become a leadership advantage.

Leaders who thrive in the AI age are not those who know the most, but those who can unlearn quickly, stay curious, and adjust how they think. They remain mentally flexible, open to new information, and willing to update their assumptions as conditions change.

At the individual level, this means developing the ability to unlearn old habits and mental models. Curiosity replaces defensiveness. The brain stays adaptable through continuous learning and reflection.

At the team level, learning shows up through fast feedback loops and safe experimentation. Teams test ideas, learn from results, and improve without waiting for perfect certainty. Mistakes become data, not failure.

At the system level, learning must travel beyond individuals and teams. Knowledge needs to move across the organization, so insights gained in one place strengthen the whole enterprise.

In the AI age, advantage no longer goes to those with the most expertise.

It belongs to those who can learn, adapt, and grow the fastest.

4. Systemic Behavior & Culture

Embedding Change That Sticks

Many AI initiatives fail not because the technology is wrong, but because the culture quietly resists it. When new behaviors are punished—through outdated incentives, hidden expectations, or unspoken norms—people revert to old ways of working. Change appears to happen, but it doesn't last.

This is why embedding change matters more than announcing it.

Lasting transformation depends on habit formation and consistent reinforcement. When behaviors are repeated daily and supported by the environment, they become natural instead of forced. Over time, change stops feeling like extra work and starts feeling like "how things are done here."

At the individual level, this begins with habits that align with identity. People act differently when new behaviors feel true to who they are, not imposed from above.

At the team level, change takes root through shared rituals and norms. Regular ways of meeting, deciding, and reflecting reinforce what matters and signal what is safe to try.

At the system level, incentives, feedback loops, and adoption patterns determine whether change spreads or stalls. Systems either reward learning and adaptation—or quietly discourage them.

Culture is not shaped by slogans or speeches.

It is shaped by what systems make easy, normal, and safe to do every day.

5. Ethical & Decision Intelligence

Wise, Context-Aware Choices

AI allows decisions to be made faster and at a much larger scale. But when decisions scale, so do mistakes. A small error, once amplified by AI, can quickly affect customers, employees, and entire systems. In this environment, ethics can no longer remain theoretical or abstract.

This is why wise, context-aware judgment matters.

Leaders must be able to slow down just enough to consider consequences, even when technology encourages speed. They need the ability to recognize bias, listen to intuition, and understand the human impact behind data-driven recommendations.

At the individual level, this means developing awareness of personal assumptions and blind spots. Leaders learn to question not just what the AI suggests, but why it suggests it—and what might be missing.

At the team level, accountability must be shared and visible. Decisions influenced by AI are discussed openly, and responsibility is not pushed onto "the system." People remain owners of outcomes.

At the system level, governance and long-term impact guide how AI is used. Rules, review mechanisms, and ethical boundaries help ensure that speed does not override responsibility.

AI can support and inform decisions.

Only humans can take responsibility for their consequences.

6. Intercultural & Inclusion Intelligence

Leading Across Differences

AI systems reflect the people, data, and assumptions behind them. Without intention, they can magnify cultural bias, reinforce exclusion, and deepen divides—especially in global organizations. This makes leadership across differences more important than ever.

Leading well in the AI age requires cultural awareness and inclusion at scale.

At the individual level, leaders begin by understanding their own identity and recognizing personal biases. They learn how background, culture, and experience shape how they interpret information and make decisions.

At the team level, inclusive leadership creates space for open dialogue. Different perspectives are welcomed, conflict is navigated with respect, and decisions are strengthened by diversity rather than weakened by it.

At the system level, global operating models must account for cultural differences instead of assuming uniform behavior. Policies, workflows, and AI systems are designed to work across regions, values, and contexts.

AI does not erase culture or bias.

It amplifies whatever is already present.

That is why inclusive, culturally aware leadership is not optional—it is essential for trust and effectiveness at scale.

7. Human–AI Synergy

Working With Intelligence Without Losing Humanity

As AI becomes part of everyday work, organizations face a choice. They can defer blindly to AI, or they can resist it out of fear. Neither approach works. The future belongs to organizations that learn how to **collaborate** with intelligence—using AI as a partner, not a replacement.

This kind of collaboration requires clarity in judgment and thoughtful redesign of how work gets done.

At the individual level, leaders learn to use AI to support their thinking without surrendering responsibility. AI assists with analysis and pattern recognition, while humans provide context, values, and final judgment.

At the team level, new models of collaboration emerge. Tasks are deliberately divided between humans and AI. Repetitive work is delegated to machines, while people focus on decisions, creativity, and exceptions that require human insight.

At the system level, enterprises redesign workflows so intelligence flows smoothly across functions. AI becomes embedded in operations, not as a separate tool, but as part of how work happens.

AI can accelerate thinking and execution.

Humans provide meaning, direction, and responsibility.

When both work together intentionally, organizations gain speed without losing their soul.

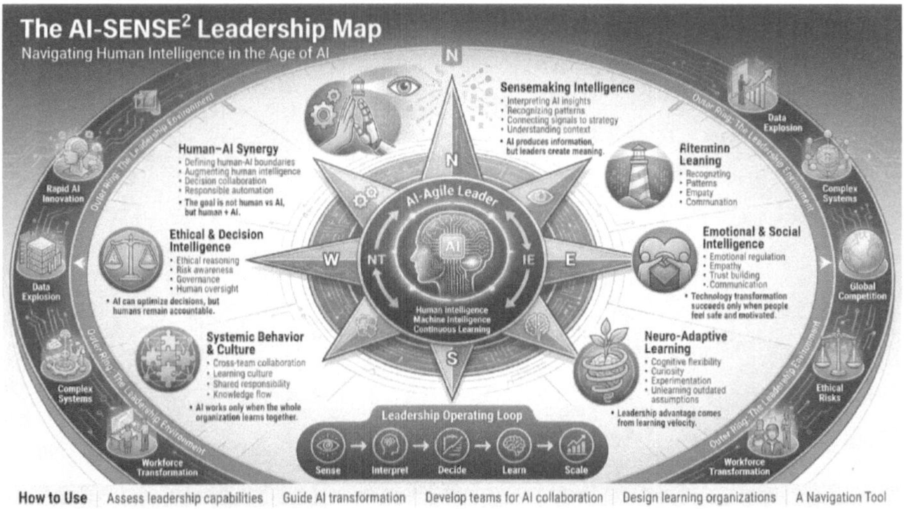

The Integrated AI-SENSE² Leader

The goal is not mastery of individual pillars.

The goal is **integration**.

An AI-SENSE² leader:

- Thinks clearly under uncertainty
- Regulates emotion under pressure
- Learns faster than disruption
- Embeds behavior through systems
- Decides ethically at scale
- Leads across cultures and machines

This leader does not chase certainty.

They build **adaptive capacity**—in themselves, their teams, and their organizations.

AI-SENSE² 7 Pillars

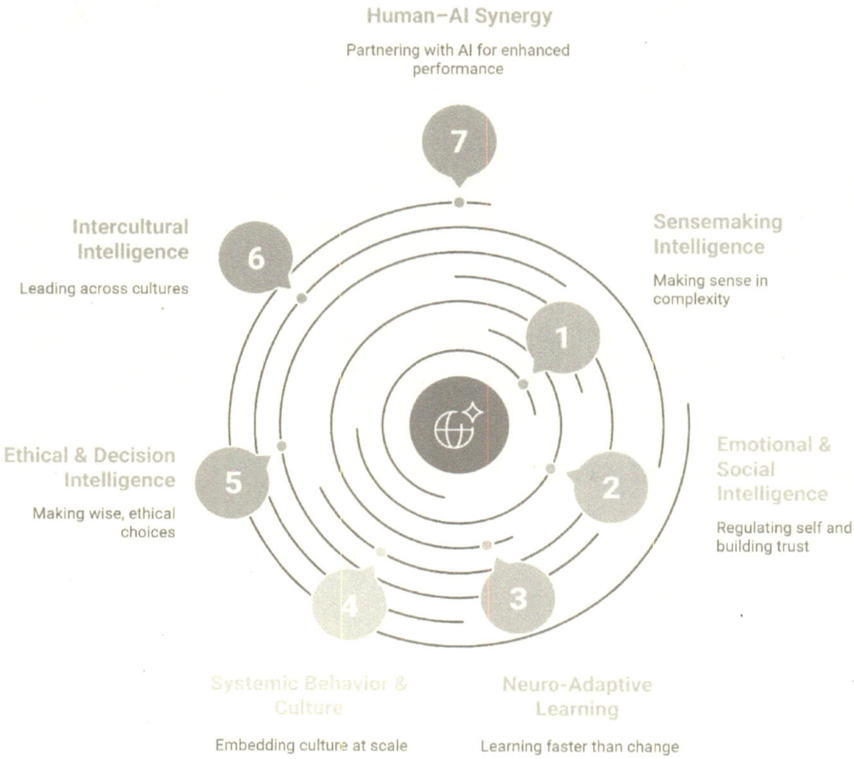

Why This Framework Works

AI-SENSE² is grounded in:

- Neuroscience (*how humans actually function*)
- Behavioral science (*how habits form*)
- Organizational research (*why change fails*)
- AI governance realities (*where accountability sits*)
- Enterprise transformation experience (*what scales and what doesn't*)

It is not aspirational.

It is **operational**.

A Final Orientation

AI will keep accelerating.

Leadership must become:

- More grounded, not more frantic
- More reflective, not slower
- More human, not less

AI-SENSE² exists to ensure that as intelligence scales, **wisdom does too**.

The chapters that follow are not theoretical explorations.

They are **capability-building journeys**—designed to help you lead with clarity, courage, and coherence in an age where intelligence is everywhere, but leadership is rare.

And that is where your work now begins.

How This Book Helps Leaders Overcome the AI-SENSE² Challenges

This book is **not a theory of AI leadership**.

It is a **practical operating guide for human adaptation in the age of AI**.

Every, challenge organizations face with AI-SENSE² exists because **leaders were never trained for this environment**. This book closes that gap—systematically.

1. It Rewires Leadership Identity (Before It Breaks)

A deeper challenge this book addresses is the quiet loss many leaders experience as their hard-earned expertise begins to feel less relevant in the face of rapid technological change. This often shows up as fear of becoming obsolete, a strong attachment to past success, or defensive leadership behaviors that protect identity rather than enable progress.

The book helps leaders reframe what leadership truly means in this new context. Instead of defining leadership as having all the answers, it shifts the focus to sensemaking—the ability to interpret signals, ask better questions, and guide others through uncertainty. Uncertainty is normalized as a leadership capability, not a personal weakness. Through real, human stories of leaders like Sarah, Arjun, and Anika, the book gently lowers psychological resistance and makes this shift feel relatable rather than threatening.

Most importantly, it helps leaders separate their self-worth from their expertise. As that separation becomes possible, leaders no longer feel the need to defend what worked in the past. They regain the freedom to experiment, learn, and design the future with confidence and humility.

2. It Builds Emotional Regulation Under AI Speed

The challenge this book addresses is not simply technological complexity, but the human cost that comes with it. Burnout, constant reactivity, and anxiety-driven decisions have become common as leaders try to keep pace with accelerating AI-driven change.

The book responds by introducing small but powerful micro-practices—simple pauses, moments of reflection, and better ways of framing questions—that help leaders slow down without falling behind. It reveals how stress is often transmitted unintentionally from leaders to their teams, shaping behavior and decision-making in subtle ways. By offering language and approaches for calm authority, even in the absence of certainty, the book helps leaders lead with steadiness rather than pressure. Over time, it replaces an addiction to urgency with what the book calls reflective speed—the ability to move forward thoughtfully and decisively at the same time.

As a result, leaders stop amplifying the chaos that AI can bring into organizations. Instead, they become stabilizing forces—creating clarity, confidence, and resilience in environments where intelligence is accelerating faster than ever.

3. It Turns Learning into a Daily Operating Habit

One of the central challenges this book addresses is slow learning velocity. Many organizations invest heavily in training, yet still struggle with fatigue, surface-level adoption, and deep resistance to unlearning old habits. Learning becomes something people attend rather than something they live, and progress stalls just when adaptability matters most.

The book shifts this pattern by reframing learning as an embedded, everyday behavior instead of a one-time event. Rather than teaching leaders and teams to learn *about* AI in abstract terms, it shows them how to learn *with* AI—using it as a partner in thinking, experimentation, and sense-making. Learning is grounded in simple, repeatable neuro-adaptive loops: trying something small, observing feedback, adjusting thoughtfully, and repeating the cycle until insight compounds.

Mistakes are no longer treated as failures that must be hidden or explained away. They are redefined as data—valuable signals that guide better decisions and faster improvement. This change alone reduces fear and unlocks curiosity.

The result is a fundamental shift in capability. Organizations stop falling behind the pace of change and begin learning faster than AI itself evolves, turning adaptability into a lasting advantage rather than a constant struggle.

4. It Replaces Silo Politics with Shared Sensemaking

Another critical challenge this book addresses is the growing tension that AI introduces across organizations. Cross-functional conflict, shadow AI initiatives, and quiet power struggles often emerge when different groups—business, technology, risk, and HR—operate with different assumptions and incentives. Decisions become fragmented, and progress turns political instead of productive.

The book helps by introducing a shared language that allows these groups to think and speak together rather than pass one another. Leaders learn how to surface hidden assumptions, align competing narratives, and recognize how informal power actually operates within their organizations. Instead of forcing agreement or defending positions, they are taught how to guide conversations toward understanding what is truly changing and why it matters.

As a result, discussions shift away from debates about who is right and toward a clearer focus on what is evolving. AI-related decisions stop being driven by politics or turf protection and become collective, grounded choices that the organization can stand behind together.

5. It Makes Ethics a Daily Leadership Muscle

A major challenge this book confronts is the way ethics is often treated as performance rather than practice. In many organizations, ethical guidelines exist on paper, yet real risks surface late—sometimes only after an AI-driven crisis has already begun. Ethics becomes policy theatre, activated too slowly to prevent harm.

The book helps leaders see that ethical failures rarely appear suddenly. They emerge quietly, long before formal violations occur. It teaches leaders how to notice early bias signals, how to ask moral questions at the start of design and decision-making, and how to intentionally create human-in-the-loop systems where judgment remains present even when outcomes are uncertain. Ethics is no longer framed as a compliance checklist, but as an ongoing exercise of responsibility and discernment in conditions where answers are not always clear.

As a result, responsible AI stops being something added after the fact. It becomes built into how decisions are made, how systems are designed, and how leaders think—ensuring that ethics is not bolted on when it is too late but woven into the work from the beginning.

6. It Converts Culture from Intent to Infrastructure

One of the persistent challenges this book addresses is why so many cultural initiatives fail to take root. Leaders make speeches, launch programs, and announce values, yet day-to-day behavior remains unchanged. Culture efforts fade because they are treated as messages rather than systems.

The book explains that culture does not change through words alone. It follows the structures people work within every day. Leaders are shown how to redesign rituals, adjust incentives, and thoughtfully embed AI into everyday workflows so that new behaviors become the natural path of least resistance. Instead of asking people to "believe differently," the focus shifts to helping them work differently.

By showing how small, consistent behaviors scale across teams and functions, the book demonstrates how culture actually forms and sustains itself. As a result, AI adoption stops feeling like another initiative layered on top of existing work and becomes simply how work gets done.

7. It Clarifies Human–AI Boundaries Before They Break Trust

Another challenge this book takes on is the confusion around trust and accountability in AI-driven decisions. In many organizations, leaders either place too much trust in AI systems or too little. Some decisions are blindly accepted because a model produced them, while others are endlessly questioned because no one is sure who is truly responsible. Accountability becomes blurred, and confidence erodes.

The book helps resolve this by introducing a clear way of thinking about decision authority. AI is positioned as an advisor that provides insight and recommendations. Humans remain the decision-makers, responsible for judgment and consequences. Systems, in turn, are designed to learn from outcomes over time. This clarity removes ambiguity about who does what and when.

Leaders are also taught how to recognize moments when AI should pause—when uncertainty is high, when signals conflict, or when ethical considerations demand deeper human involvement. At every level, human accountability is reinforced, not diluted.

The result is a healthier form of trust. Confidence in AI increases, but control is not surrendered. Decisions become more reliable because responsibility is clear, shared appropriately, and consciously owned.

8. It Prepares Leaders for Scale—Not Just Pilots

A common frustration this book addresses is why AI often succeeds in labs and pilot projects yet fails to deliver real value at the enterprise level. Proofs of concept look promising, but when organizations try to scale them across teams, regions, and business units, progress stalls and impact fades.

The book draws on real enterprise cases from industries such as logistics, aerospace, healthcare, and manufacturing to show what actually breaks during scale-up—and how to fix it. It explains how the **AI-SENSE²** approach works across different geographies, organizational cultures, and regulatory environments, without losing coherence or control. Rather than treating strategy, execution, and governance as separate efforts, the book connects them into a single, practical flow.

As a result, AI stops living in isolated experiments and begins to generate sustained business advantage. What once worked only in controlled environments becomes embedded in how the enterprise operates, decides, and competes at scale.

9. It Protects Leaders from Cognitive Overload

This book also addresses a growing but often unspoken challenge facing leaders today: decision fatigue, loss of focus, and burnout. As AI increases the volume and speed of information, leaders are expected to process more signals, make more choices, and respond more quickly than ever before.

The book helps leaders understand what kinds of thinking can be safely delegated to AI and what must remain deeply human. Routine analysis and pattern recognition can be offloaded, while attention is intentionally protected for judgment, ethics, and meaning. By making this distinction clear, leaders are able to conserve their cognitive energy for the decisions that truly matter.

In doing so, the book introduces a more sustainable way of leading. Leadership is no longer about pushing harder or moving faster at all costs, but about designing work and decision-making in ways that can be sustained over time. The result is leaders who remain effective longer—not just quicker in the moment, but steadier and wiser across the long arc of change.

10. It Gives Leaders a Coherent Operating System

One of the final challenges this book addresses is the confusion created by fragmented frameworks and constant tool overload. Leaders are exposed to countless models, methods, and AI tools, each promising clarity but often adding to the noise. Instead of insight, many experience fatigue and uncertainty about what actually matters.

The book cuts through this complexity by integrating neuroscience, leadership practice, human–AI collaboration, and organizational culture into a single, coherent approach. Rather than asking leaders to juggle disconnected ideas, it introduces one memorable and practical model: **AI-SENSE^2**. This model is designed to work across levels—from the self, to teams, to the broader system—so leaders can apply it consistently without constant reinvention.

As a result, leaders do more than simply adopt AI tools or frameworks. They develop the clarity, judgment, and coherence that their organizations are missing. In a world filled with intelligence, they become the intelligence their organizations truly need.

The True Promise of This Book

This book does not promise:

- Faster AI
- Smarter tools
- Perfect decisions

It promises something more lasting:

Leaders who can think clearly, learn continuously, make wise decisions, and lead both humans and machines—without breaking themselves or their organizations.

AI-SENSE^2 Manifesto

Intelligence Without Wisdom Is Dangerous.

We are not entering the age of artificial intelligence.

We are entering the age of amplified intelligence.

And amplification does not fix weakness.

It magnifies it.

If leadership is confused, AI accelerates confusion.

If culture is fragile, AI exposes fragility.

If decisions lack ethics, AI scales damage.

The real risk is not machines becoming powerful.

The real risk is humans remaining unprepared.

AI-SENSE² Exists for One Reason:

To ensure human wisdom grows as fast as machine intelligence.

We Believe

Intelligence without sensemaking becomes noise.

Intelligence without emotional regulation becomes fear.

Intelligence without learning velocity becomes irrelevant.

Intelligence without systems thinking becomes chaos.

Intelligence without ethics becomes harm.

Intelligence without intercultural fluency becomes division.

AI does not solve these problems.

It intensifies them.

AI-SENSE² Is Not a Framework for Technology.

It is a framework for upgrading human capacity.

Before scaling AI, we must scale:

- Clarity under ambiguity
- Emotional stability under pressure
- Learning speed under change
- Structural alignment across systems
- Ethical reasoning under uncertainty
- Collaboration across cultures and machines

SENSE Before Scale

- Sensemaking before automation.
- Emotional regulation before acceleration.

- Neuro-adaptation before disruption.
- Systems alignment before expansion.
- Ethical intelligence before optimization.

Only then does the squared intelligence emerge.

The Squared Reality

We are no longer operating human-to-human.

We are operating:

Human × Human

Human × Machine

Culture × Culture

System × System

This is intelligence multiplied.

If leaders cannot coordinate multiplied intelligence, the system fractures.

This Is a Call to Leaders

Do not adopt AI faster than you evolve.

Do not automate what you do not understand.

Do not scale what you cannot govern.

Do not optimize what you cannot ethically defend.

The Future Belongs to Those Who:

Learn faster than change.

Stay calm under complexity.

Think systemically, not reactively.

Choose wisely under uncertainty.

Build cultures that can metabolize intelligence.

AI-SENSE² In One Line

It is the discipline of ensuring that as intelligence increases, coherence increases.

Because when intelligence rises without coherence, collapse follows.

When intelligence rises with coherence, civilization advances.

Why AI-SENSE² Is a Trusted and Authentic Leadership Framework

AI-SENSE² is grounded in a simple but research-validated reality:

AI does not fail because it lacks intelligence—AI fails because leadership systems are not ready to guide intelligence responsibly.

Across industries, academic research and enterprise studies consistently show that the limiting factor in AI success is **human capability, judgment, and leadership maturity**, not technology.

1. AI Adoption Is High—But AI Impact Remains Low

Global research on large companies shows that **AI is now everywhere**.

About **88 out of 100 organizations already use AI** in at least one part of their business. Starting with AI is no longer difficult. Many teams can build a model, buy a tool, or run a pilot.

But when we look deeper, a big gap appears.

Only **about 20 to 25 out of 100 companies are able to scale AI** across the whole organization. Even fewer—**less than 30 out of 100—can clearly show real business results** from their AI efforts.

This tells us something important.

Using AI is easy. Leading with AI is rare.

Most organizations struggle not because of technology, but because they lack the leadership skills needed to grow AI in a healthy and meaningful way.

Scaling AI requires people who can make sense of complex situations, align teams around a clear purpose, make ethical decisions under pressure, and build systems that learn and improve over time.

These are exactly the leadership capabilities that **AI-SENSE²** is designed to develop.

2. Leadership, Not Algorithms, Is the Primary Risk Factor

Research shows something worrying about how companies use AI today.

Almost **95 out of 100** leaders say their company has already faced some kind of AI problem or failure. But at the same time, only about 2 out of 100 companies are truly ready to use AI in a responsible and safe way.

This tells us an important thing:

Most AI problems are not caused by bad technology.

In many cases, the AI works exactly as it was built to work.

The real problems come from human decisions.

Often, it is not clear who is responsible for AI decisions.

Humans do not properly check or question what the AI suggests.

People trust AI outputs too much, even when they feel something is wrong.

When things fail, no one clearly owns the mistake.

So, what looks like an AI failure is usually a leadership and decision failure.

This is where AI-SENSE² helps.

AI-SENSE² treats ethics and responsibility not as technical rules, but as everyday leadership habits. It teaches leaders and teams how to think clearly, stay involved, and take responsibility when working with AI.

In this approach, responsible AI does not come only from tools or controls.

It comes from aware leaders, clear decisions, and strong human judgment.

3. Human Judgment Improves AI Outcomes—Research Confirms It

Studies on how humans and AI work together show an important lesson.

When decisions are complex or unclear, teams where humans and AI work together perform better than teams that rely only on AI or only on people. AI is powerful, but it does not understand context the way humans do. When humans add real-world understanding, intuition, and proper framing, AI decisions become more accurate and useful.

The problem starts when people trust automation blindly. When AI outputs are followed without questioning—especially in uncertain situations—decision errors actually increase. This happens because AI cannot fully understand human values, emotions, or unexpected situations on its own.

These findings support a key idea behind AI-SENSE².

AI is meant to support human judgment, not replace it. The best results come when people stay actively involved—thinking, questioning, and guiding the AI.

In this new world, leaders must learn how to think with AI, not switch off their thinking because AI is present. Good leadership means using AI as a partner in decision-making, while keeping human responsibility firmly in place.

4. Emotional Regulation and Psychological Safety Are Proven Performance Drivers

Research from neuroscience and organizational psychology shows something very important about how people make decisions.

When people are mentally overloaded—too much information, too many alerts, too many choices—their decision **quality drops by 30 to 40 percent.** Even smart and experienced leaders start making weaker decisions when their minds are overwhelmed.

Stress makes the problem worse. When leaders are under pressure and cannot manage their emotions, they are more likely to take unnecessary risks or blindly follow AI recommendations without questioning them. This increases automation bias, where people trust AI even when it may be wrong.

Another key finding is about psychological safety. Teams that feel safe to speak up, question decisions, and admit mistakes learn faster and adapt better. Psychological safety is one of the strongest signs of high performing, learning teams.

This is why AI-SENSE^2 focuses not only on technology, but also on emotional and social intelligence. Intelligent systems do not operate in isolation—they amplify the mental and emotional state of the humans using them.

When leaders are calm, aware, and grounded, AI outcomes are more stable and reliable. But when leaders are stressed and unregulated, even the best data and models can lead to unstable and risky results.

Research in neuroscience and psychology shows that when people are mentally overloaded, their decision **quality drops by 30–40%.** Stress and poor emotional control make leaders more likely to take risks or blindly trust AI, increasing mistakes.

Psychological safety plays a key role in learning and adaptation. Because AI systems amplify human emotions and behavior, AI-SENSE^2 focuses on emotional and social intelligence. Unregulated leaders can

5. Learning Velocity Matters More Than Expertise

Research on learning shows that skills do not last as long as they used to. Both technical and leadership knowledge become outdated quickly. Traditional, one-time training does not work well in fast-changing environments.

What works better is continuous learning—where people learn, apply, reflect, and adjust again and again. Teams that follow this learning loop adapt faster than those that rely on fixed skills.

This is why AI-SENSE² focuses on neuro-adaptive learning. It shifts leadership from *what you know* to how quickly you can learn and adapt.

6. Culture and Coaching Are the Hidden Scaling Mechanisms

Studies on enterprise transformation show a clear pattern. Technology initiatives often fail when leadership behavior stays the same. Even the best AI tools cannot succeed in a culture that resists change.

Whether AI becomes part of everyday work—or gets quietly abandoned—depends largely on culture. Research also shows that **coaching changes behavior faster and more deeply** than rules, policies, or one-time training.

This is why **AI-SENSE²** includes coaching, culture, and system-wide behavior. Intelligence can scale only when **daily habits and ways of working scale with it**.

Why This Makes AI-SENSE² Credible and Trustworthy

AI-SENSE² is trusted because it is:

- **Research-aligned**, not trend-driven
- **Human-centred**, not tool-obsessed
- **Behavioral**, not theoretical
- **Integrative**, not fragmented
- **Tested across leadership, coaching, innovation, and enterprise transformation**

It does not promise control over AI.

It prepares leaders for **responsibility in the presence of intelligence**.

Final Proof Point

The evidence is clear:

When intelligence accelerates, leadership must deepen.

AI-SENSE^2 exists because research, practice, and lived experience all point to the same conclusion:

The future of AI success depends less on smarter machines—and more on wiser leaders.

Part I

The Leadership Shift AI Forces

Introduction

Why Fast Machines Are Slowing Down Great Leaders

Artificial Intelligence (AI) did not come slowly. It came very fast and with a big promise: to help people work better and faster.

Today, machines can do in seconds what used to take humans days or even weeks. They can study huge amounts of data, find hidden patterns, and predict what might happen next.

In many companies, this looked like great success.

Reports looked good.

Leaders were impressed by AI demonstrations.

Productivity numbers went up.

But inside many organizations, something still felt wrong.

Even with powerful AI tools:

- Decisions were still slow
- People felt worried instead of confident
- Leaders worked longer hours but felt less in control

Teams started using AI tools, but the way people worked did not really change. In some cases, company performance even became worse.

What was supposed to make organizations faster started to show deeper problems.

This chapter is about those problems.

When New Technology Meets Old Systems

The stories in this chapter are not about bad technology.

They are about **good technology meeting old systems that were never designed for it**.

This shows an important truth about the AI age:

AI success is not only a technology problem.

It is a **leadership and human capability problem**.

In the following pages, you will see leaders who learned that experience and logic alone are not enough anymore.

They discovered a new way of thinking.

I call this **the Third Thinking System**.

The best leaders today are not the ones who know everything.

They are the ones who know how to **work with both human thinking and machine intelligence**.

They learn when to trust AI and when to use their own judgment.

The Hidden Problem with Speed

AI makes everything faster.

But speed alone does not create success.

When organizations move fast without understanding the situation clearly, they can make poor decisions that create problems later.

Some companies rushed to adopt AI because they were afraid of falling behind their competitors.

But these early decisions sometimes created systems that were hard to change later.

True success with AI comes from three things:

Clear Purpose

Understanding why you are using AI.

Right Decisions

Choosing the right kind of intelligence for the problem.

Better Systems

Changing how the organization works instead of just adding new tools to old processes.

The Personal Challenge for Leaders

AI is also changing how leaders feel about their work.

Many leaders built their careers on knowledge and experience.

But AI can now analyze information faster than any human.

This can make leaders feel uncertain.

In this fast-moving world, leaders often feel pressure to make quick decisions. But if they move too fast without thinking carefully, mistakes happen.

At the same time, if they spend too much time thinking and not enough time acting, they fall behind.

This balance between **speed and reflection** becomes very important.

Coaching and learning become essential, because leaders must develop new skills that AI cannot replace.

A Chance to Grow

In many organizations today, AI systems can analyze data in seconds, but human decisions still take weeks.

Some companies are now changing how they lead and make decisions.

They are moving from simply checking work to using better judgment.

They are redesigning their culture, rewards, and rules so humans and AI can work together.

These changes show an important lesson:

AI is not slowing organizations down.

Old leadership habits are.

This chapter invites you to rethink how leadership should work in a world where intelligence is everywhere, speed is unavoidable, and responsibility still belongs to humans.

AI has already changed organizations.

The real question is:

Are leaders ready to change as well?

Story: 1

The Third Thinking System: Why AI Agility Is the New Leadership Skill

> **"A wealth of information creates a poverty of attention."**
>
> **– Herbert A. Simon:**

Imagine you're driving home on a rainy evening. The road is quiet, your thoughts drifting, when suddenly a dog darts onto the street. Your foot slams the brake before your mind has time to form a sentence. Later that same night, you're sitting at your desk, sketching a strategy for your team—thinking through risks, trade-offs, timelines, and long-term impact. You pause, reconsider, erase, and start again.

These two moments feel completely different, yet they come from the same place: the human mind. They reveal how we think in different modes depending on what the moment demands. And today, something new has entered this mental system—a third way of thinking that is reshaping how decisions are made. This is where AI Agility begins.

For a long time, our thinking lived in **two lanes**.

One lane is fast. It's instinctive, automatic, and emotional. It reacts before we are aware we're reacting. This is the part of the mind that keeps us safe, reads a room instantly, finishes other people's sentences, and jumps to conclusions. It saves time and energy—but it can also be wrong.

The second lane is slow. This is deep, deliberate thinking. It analyzes, calculates, questions, and reasons. It helps us solvan e complex problems and make thoughtful decisions. But it comes at a cost. It takes time, focus, and mental energy, and it gets tired.

For centuries, every human decision was shaped by the tension between these two modes: speed versus depth, instinct versus logic.

Now, **a third mode** has entered the picture.

AI thinking is not biological and not emotional, but it is extraordinarily powerful. It can process vast amounts of data in seconds, spot patterns humans would never notice, and combine speed with precision in a way neither instinct nor logic can manage alone. It doesn't get tired. It doesn't have ego. It doesn't carry emotional baggage or cognitive fatigue.

But AI is not here to replace human thinking. In the context of AI Agility, it exists to extend it.

AI Agility is the ability to move smoothly between fast thinking, deep thinking, and AI-assisted thinking—using each at the right moment. In fast-moving situations, AI can act like a heightened sense of awareness, flagging risks or opportunities before they become obvious. In complex situations, it can take on the heavy cognitive load—summarizing information, exploring options, and connecting patterns—so humans can focus on meaning, judgment, creativity, and ethics.

When used well, AI doesn't shrink human intelligence. It expands it. Individuals become stronger thinkers. Teams gain clarity faster. Decisions become more informed, timelier, and more forward-looking.

Yet this power comes with a real risk.

The danger is not AI itself, but the temptation to stop thinking. When leaders defer judgment entirely to machines, their own reasoning weakens. Critical thinking fades. Bias goes unchallenged. Decisions may be fast, but they become shallow.

AI Agility requires something different. It asks leaders to stay in the driver's seat. To direct AI, not surrender to it. The human remains responsible for judgment, values, and consequences. AI becomes the engine that increases reach and speed, not the authority that replaces responsibility.

This is the true promise of AI Agility: a partnership where human intuition, human reasoning, and machine intelligence work together, each doing what it does best.

The future will belong to leaders who can move fluidly between these three modes. Not the ones who know the most, but the ones who can shift how they think, guide AI with wisdom, and combine human insight with machine intelligence.

This is the new frontier of thinking. This is AI Agility. And it is changing how we decide, lead, and shape what comes next.

Here are the top 3 key takeaways:

- Leadership now involves mastering three types of thinking. Beyond quick instinct and slow logic, leaders must learn to work with AI as a third thinking system that enhances human abilities rather than replacing them.
- AI should strengthen human judgment, not replace it. True AI agility involves using AI thoughtfully to improve insight and analysis, while humans stay responsible for meaning, ethics, and decision-making.
- The leaders who succeed are those who can shift their thinking styles easily. Success in the AI era goes to leaders who understand when to rely on intuition, when to think deeply, and when to engage AI—without losing control.

Story: 2

Why AI Success Is a Human Capability Problem, not a Technology Problem

> **"In times of change, learners inherit the earth, while the learned find themselves beautifully equipped to deal with a world that no longer exists."**
>
> **– Eric Hoffer**

Organizations that don't manage the human adaptation cost eventually pay it in attrition, performance loss, or failure to scale.

Most leaders today are not struggling with a lack of ambition, intelligence, or technology. They are struggling with **speed**.

AI is advancing faster than organizational learning cycles. Teams are expected to adapt continuously, collaborate across cultures and time zones, make ethical decisions with AI-generated insights, and stay emotionally resilient in an environment of permanent uncertainty.

This is not a technology problem. It is a **human adaptation problem**.

And this is where most transformation efforts quietly fail.

We train tools, deploy platforms, and announce change programs—while assuming people will somehow "catch up." They don't. Instead, anxiety rises, learning slows, trust erodes, and adoption becomes performative rather than real.

AI-SENSE^2 was born from observing these failures repeatedly across enterprise transformations—not from theory, but from what breaks in real organizations.

Sensemaking Before Speed

One of the most underestimated leadership skills today is sensemaking. Teams don't resist change because they are stubborn; they resist because change often arrives without meaning. When AI initiatives are framed only in terms of efficiency or cost, people fill the gaps with fear.

Sensemaking intelligence allows leaders to interpret ambiguity, integrate multiple perspectives, and create shared understanding. This capability is deeply grounded in organizational research showing that humans tolerate uncertainty far better than confusion. When leaders explain *why* change is happening—not just *what*—resistance drops, alignment improves, and energy returns.

This is not soft leadership. It is cognitive load management at scale.

Emotional Intelligence Is No Longer Optional

AI accelerates work faster than human nervous systems evolved to handle. Under constant pressure, people don't become more productive; they become reactive. Burnout, disengagement, and silent quitting are not individual failures—they are predictable biological responses to unmanaged stress.

AI-SENSE² treats emotional and social intelligence as operational capabilities, not personality traits. Decades of neuroscience and large-scale studies, including Google's Project Aristotle, consistently show that psychological safety and emotional regulation are prerequisites for high performance. Leaders who can regulate themselves, read team dynamics, and build trust create environments where adaptation is possible.

Ignoring emotion doesn't make leaders strong. It makes systems fragile.

Learning Velocity Beats Static Expertise

The half-life of skills is shrinking. Training programs designed for stable environments are no longer enough. What differentiates future-ready teams is not what they know, but how fast they can learn, unlearn, and relearn.

Neuro-adaptive learning, grounded in neuroplasticity research, focuses on building curiosity, cognitive flexibility, and habit-based learning into daily work. Organizations that prioritize learning velocity consistently outperform peers in innovation and resilience. Those that cling to static expertise fall behind—even when they hire "top talent."

The future belongs to teams that learn faster than change, not those who mastered yesterday's tools.

Culture is a System, not a Slogan

Most leaders understand culture matters. Fewer understand why culture change fails.

Culture doesn't shift because of town halls or value statements. It shifts when behaviors are reinforced—or punished—by systems. Incentives, workflows, metrics, and rituals quietly shape what people actually do.

AI-SENSE² applies behavioral science to embed new ways of working into the fabric of organizations. When systems support desired behaviors, change sticks. When they don't, even the most motivated teams revert under pressure.

Culture is not what leaders say. It is what the system makes easy.

Ethics Is Now a Leadership Competency

AI can process information at scale. It cannot have consequences.

As AI increasingly informs decisions, leaders must navigate ambiguity, bias, and accountability with greater clarity. Ethical and decision intelligence ensures humans remain meaningfully "in the loop," balancing speed with responsibility.

This is not theoretical. AI governance failures are already costing organizations billions in fines, lawsuits, and reputational damage. Regulators are clear: accountability rests with people, not algorithms.

In the AI era, judgment—not intelligence—is the real differentiator.

Leading Across Cultures and Machines

Work is now global, digital, and increasingly human–AI hybrid. Teams collaborate across cultures, time zones, and intelligent systems. Without intentional design, this complexity amplifies misunderstanding and exclusion.

AI-SENSE² integrates intercultural leadership with human–AI collaboration, helping leaders redesign roles, workflows, and norms so that humans and machines complement rather than compete with each other. Research consistently shows that inclusive, culturally intelligent teams outperform homogeneous ones in innovation and adaptability.

AI doesn't neutralize culture. It magnifies it.

Why This Framework Is Worth Trusting

AI-SENSE^2 is not a prediction of the future. It is a response to what is already happening.

It draws from neuroscience, behavioural science, organizational research, and hard-earned lessons from enterprise transformation failures. It focuses on capabilities leaders can develop—not traits they must be born with. And it is designed to scale across systems, not rely on heroic individuals.

In a world where technology is advancing faster than institutions, leaders don't just need better tools. They need a way to help humans adapt without breaking.

That is what AI-SENSE^2 is built to do.

It is a **response to patterns that have already repeated across decades**, now amplified by AI speed.

It integrates:

- Neuroscience (*how humans function*)
- Behavioural science (*how habits form*)
- Organizational research (*why change fails*)
- AI governance realities (*where accountability sits*)
- Enterprise transformation data (*what scales and what doesn't*)

Here are the **3 key learnings**:

- **AI fails when humans can't adapt fast enough.**

  The real bottleneck isn't technology—it's sensemaking, emotional resilience, learning speed, and trust. AI-SENSE^2 exists to close this human adaptation gap.

- **Speed without meaning creates fear, not agility.**

  Leaders must help people understand *why* AI change is happening, regulate stress, and build learning into daily work. Sensemaking and emotional intelligence are now operational necessities.

- **Judgment, learning velocity, and culture decide success.**

  AI-SENSE^2 focuses on building human judgment, ethical accountability, and system-level culture so organizations scale AI responsibly instead of breaking under it.

Story: 3

Why Brilliant Leaders Struggle in the AI Age (And What Coaching Fixes)

> **"The illiterate of the 21st century will not be those who cannot read and write, but those who cannot learn, unlearn, and relearn."**
>
> **– Alvin Toffler**

Dr. James Rodriguez had done everything right.

As Vice President of Engineering at a Fortune 500 company, he was respected for his technical depth, strategic rigor, and flawless execution. For seven years, his expertise made him indispensable. He was the person teams turned to when problems were complex and stakes were high.

Then enterprise AI arrived—and the ground beneath him began to move.

Within six months, parts of his hard-earned expertise were no longer as valuable as they once were. Decisions that used to unfold over weeks now demanded answers in hours. AI-powered initiatives produced consequences no one—including James—had anticipated. Nearly forty percent of his team burned out or left for AI-native firms. His workweeks stretched to seventy hours, yet he never felt caught up.

At his annual review, his CEO paused and said something James still remembers clearly:

"James, you're still brilliant. But the world has changed. We need you to lead differently—faster, more adaptively, with greater emotional intelligence. Can you evolve?"

James wanted to say yes.

Instead, he froze.

He didn't need another certification or a deeper technical course. More knowledge wasn't the problem. What he lacked was *capacity*—the mental, emotional, and decision-making bandwidth to lead in a world that no longer behaved the way it used to.

That was the moment coaching entered his life.

As James began to reflect, he realized his struggle wasn't personal failure. It was systemic. The AI age creates a set of leadership crises that quietly affect almost everyone who steps into it unprepared.

The first crisis hit at the core of his identity. For years, James had defined himself as "the expert." But AI accelerated knowledge creation so quickly that static expertise began to expire. When algorithms outperformed him in certain domains, it wasn't just uncomfortable—it was destabilizing. His confidence wavered, not because he lacked ability, but because his identity was tied too tightly to having answers.

Through coaching, James learned to separate who he was from what he knew. Expertise stopped meaning "I have the answers" and started meaning "I ask better questions and apply judgment over AI outputs." His relevance shifted from execution to interpretation, ethics, and sensemaking.

The second crisis arrived through speed. AI multiplied both how fast decisions were made and how far their consequences travelled. James went from making a handful of major decisions each week to dozens, many affecting thousands of people instantly. Without reflection, speed became dangerous. With fear, it became paralyzing.

Coaching didn't slow him down. It taught him reflective speed. James adopted short, disciplined pauses before critical decisions—quick scans of consequences, values, and reversibility. These moments took minutes, not hours, but they prevented long-term damage. He learned that speed without reflection creates risk, and reflection without speed creates irrelevance. Leadership requires both.

Then came the emotional toll. AI environments were rejection heavy. Experiments failed publicly. Projects were cancelled abruptly. Data exposed mistakes instantly. James tried to power through—until his resilience collapsed.

What coaching revealed was counterintuitive: resilience wasn't toughness. It was a regulation. James learned to separate facts from feelings, interrupt emotional spirals, and treat failure as information rather than identity. He began asking himself, "This feels bad—but what is actually true?" That single question changed how long he could endure pressure without breaking.

Another tension followed: overthinking versus recklessness. AI punished hesitation and impulsiveness alike. James found himself oscillating between analysis paralysis and rushed decisions. The answer wasn't perfection—it was learning velocity. He learned to categorize decisions by reversibility, match speed to stakes, and build rapid learning loops: decide, observe, adjust. Progress replaced paralysis.

As AI dashboards multiplied, another problem surfaced—signal overload. Everything looked urgent. Metrics screamed for attention. James became reactive, chasing numbers instead of meaning. Coaching helped him step back and design filters instead of adding more dashboards. He defined a small set of true priorities, created signal tiers, and introduced grounding rituals that kept him oriented. Without that clarity, he realized, AI would decide his priorities for him.

The final crisis was quieter but deeper: a loss of purpose. AI-driven success delivered results, but something felt hollow. His team executed well, yet morale dipped. Ethical unease surfaced. Coaching helped James confront a hard truth—AI without purpose becomes extractive. Purpose wasn't a slogan; it was a decision filter. He began asking different questions: Does this make people more capable or more dependent? Does it increase freedom or manipulation? Would we be proud to explain this openly?

Those questions restored meaning.

James didn't survive the AI age because the world became easier. He survived because he grew faster than the disruption.

What coaching gave him wasn't therapy, training, or consulting. It was developmental infrastructure—a space for identity evolution, emotional regulation, reflective decision-making, accountability, and long-term capacity building.

The lesson is simple but demanding. In an AI-driven future, the leaders who thrive won't be the smartest or most technical. They will be the most grounded, reflective, emotionally resilient, and purpose driven.

You don't prepare for the AI world by learning tools alone.

You prepare by building identity beyond expertise, reflective speed, emotional resilience, learning velocity, discernment, and clarity of purpose.

Those capacities don't emerge by accident. They are built deliberately, over time—often with coaching.

The AI age doesn't need perfect leaders.

It needs human leaders who can evolve without losing themselves.

This story isn't just about James. It's a mirror—and an invitation.

The question is no longer, *"Can AI change the world?"*

It's *"Can we change fast enough—and wisely enough—to lead it?"*

Now you know what that preparation looks like.

Here are the **top 3 key takeaways for readers**:

- **AI doesn't make leaders irrelevant; clinging to old identities does.**

  Leaders must separate their self-worth from static expertise and evolve from "having answers" to exercising judgment, sensemaking, and ethical leadership.

- **Speed without reflection is dangerous; reflection without speed is useless.**

  The AI age demands *reflective speed*—the ability to make fast decisions while pausing just enough to assess impact, ethics, and reversibility.

- **Coaching builds the human capacities AI cannot replace.**

  Emotional regulation, learning velocity, purpose clarity, and discernment don't emerge naturally under pressure—they are deliberately developed, and coaching is the infrastructure that makes that growth possible.

Story: 4

Speed Without Sense: The Silent Traps of Early AI Decisions

> **"There is nothing so useless as doing efficiently that which should not be done at all."**
>
> **– Peter Drucker**

In 2022, it felt like standing at the edge of a gold rush.

Every boardroom conversation carried the same tension: *"If we don't act now, we'll be left behind."*

Leaders moved fast. Contracts were signed. Platforms were chosen. AI roadmaps were announced with pride. Speed became the strategy.

But speed, as they would soon learn, has a shadow.

The Questions Organizations Couldn't Ignore

1. **Are we moving fast because it's right—or because everyone else is?** Have we paused to understand the long-term consequences of our early AI decisions, or are we mistaking speed for strategy?
2. **Are we building real business advantage—or just impressive demos?** Do our AI investments solve meaningful

problems, or are they optimized for executive presentations and vendor case studies?

3. **How easy would it be to change our AI strategy tomorrow?** If a better model, platform, or partner emerges, can we pivot—or are we already trapped by contracts, architectures, and skills?
4. **Do we actually need this level of AI power for this problem?** Are we overengineering solutions where simpler, faster, and cheaper models would deliver better outcomes?
5. **Whose intelligence is really driving our AI—ours or the vendor's?** Are we leveraging our unique data and institutional knowledge, or outsourcing our differentiation without realizing it?
6. **Are we choosing closed systems for comfort—or open systems for control?** Have we confused perceived safety with strategic leverage, and what happens when negotiating power disappears?
7. **Is our cloud strategy serving innovation—or silently shaping it?** Are we cloud-first with intention, or cloud-dependent by default—at the cost of sovereignty, flexibility, and resilience?
8. **Does our governance accelerate learning—or kill momentum?** Are we managing AI through rigid approvals and fear, or through principles that encourage responsibility and progress?
9. **Have we prepared our people for AI—or just our infrastructure?** Do our teams feel empowered to work with AI, or anxious about relevance, trust, and role erosion?
10. **Are we treating AI as a one-time initiative—or a system that must evolve?** When the technology, ethics, and markets change—as they inevitably will—are we ready to adapt continuously?

Why These Questions Matter

These weren't theoretical debates. They were the *silent friction points* slowing adoption, eroding trust, and inflating costs—long before organizations had the language to explain what was wrong.

The learnings that follow didn't emerge from success stories. They emerged from **living through these questions—again**.

Learning 1: Speed Without Sense Creates Invisible Chains

At first, the momentum felt intoxicating. Teams deployed powerful generative AI models built by major vendors, amazed by their performance. Demos impressed executives. Productivity spiked.

Yet, quietly, something else happened.

Customization became harder. Experiments felt constrained. What once seemed like acceleration slowly turned into dependency. The organization had moved fast—but without fully sensing where the road would end.

Only later did leaders realize **speed without sense doesn't create freedom; it creates locks.**

Learning 2: Power Is Not the Same as Advantage

The most expensive models delivered breathtaking results—but not always meaningful ones. Leaders began asking uncomfortable questions.

"Are we solving real problems... or just showcasing capability?"

The truth emerged slowly: competitive advantage didn't come from using the *most powerful* AI—it came from using the *right* AI in the *right* context. Strategic clarity mattered more than raw computational force.

Learning 3: Lock-In Is a Tax You Pay Every Day

As markets evolved and better models appeared, switching became painful. Contracts, architectures, and skills were deeply tied to early choices.

What once felt like a smart partnership now felt like a toll road with no exit ramps. Leaders learned the hard way that **vendor lock-in isn't a one-time decision—it's a daily cost on innovation, negotiation, and choice.**

Learning 4: Not Every Task Needs a Supercomputer

One breakthrough moment came when teams experimented differently.

They used small, low-cost models early in development. Mid-range models for refinement. High-powered models only when precision truly mattered.

Suddenly, costs dropped. Speed increased. Decisions improved.

The insight was simple yet profound: **intelligence is not about size—it's about fit.**

Learning 5: Your Data Is Your Identity

For companies rich in original content and proprietary knowledge, a deeper realization emerged.

No external model—no matter how advanced—understood their world as they did. Their data held nuance, history, and competitive DNA.

This sparked a shift: *"Where differentiation matters, we build."*

AI stopped being just a tool and became an extension of organizational identity.

Learning 6: Open Source Isn't Risk—It's Leverage

As open-source models matured, leaders noticed something unexpected.

They created healthy pressure. They prevented complacency. They restored balance in negotiations.

Open ecosystems didn't eliminate proprietary solutions—but they ensured no single player held all the power. **Optionality became a strategic stance, not a technical preference.**

Learning 7: Cloud Is a Means, not a Mandate

Cloud adoption followed a similar arc.

Multi-cloud strategies weren't about fashion or fear—they were about focus. When infrastructure worries faded, teams redirected energy toward innovation and value creation.

The smartest leaders adopted a clear principle: *cloud-first, but not cloud-blind.*

Security, sovereignty, and control remained non-negotiable.

Learning 8: Governance Should Enable, Not Suffocate

Initially, governance slowed everything down. Approvals piled up. Innovation stalled.

Then something changed.

Organizations shifted governance from rigid rules to guiding principles—governing outcomes, ethics, and accountability rather than tools.

Suddenly, governance became a guardrail, not a roadblock.

Learning 9: Technology Changes Fast—People Don't

The hardest lesson wasn't technical.

Teams felt uncertain. Roles blurred. Trust wavered. The emotional weight of AI transformation surfaced quietly but powerfully.

Leaders who ignored this struggled. Those who listened, empathized, and created psychological safety saw adoption flourish.

AI succeeded where humans felt safe to learn, fail, and evolve.

Learning 10: AI Is Not a Project—It's a Living System

By the end, the biggest realization was unmistakable.

AI wasn't something you *implement* and move on from. It was something you *grow with*. Models evolve. Markets shift. Ethics deepen. Cultures adapt.

Organizations that treated AI as a living capability—not a checklist—began to compound learning, value, and resilience.

The Quiet Conclusion

The AI gold rush didn't end.

It matured.

The winners weren't the fastest movers—but the deepest learners. They built systems that could think, adapt, and choose wisely. They preserved optionality. They respected humans as much as machines.

And in doing so, they didn't just adopt AI.

They learned how to evolve with it.

Here are the **top 3 key takeaways**:

- **Speed without sense creates long-term lock-in.**

  Moving fast with AI feels powerful, but without careful sensemaking, it leads to dependency, reduced flexibility, and costly vendor and architectural traps.

- **Advantage comes from fit, not raw AI power.**

  The most successful organizations use the *right* level of AI for the *right* problem, leverage their own data as differentiation, and avoid overengineering for show.

- **AI succeeds when humans are designed into the system.**

  Governance must enable learning, people must feel psychologically safe, and AI must be treated as a living capability—not a one-time project—so organizations can adapt continuously.

Story: 5

When AI Thinks in Seconds but Organizations Decide in Weeks

> **"A bad system will beat a good person every time."**
>
> **– W. Edwards Deming**

Ravi stood in the executive conference room, staring at a dashboard that made no sense.

The numbers were beautiful. Green everywhere. Cycle times are down. Forecast accuracy up. Response times cut in half.

And yet—nothing felt faster.

Three months earlier, the organization had proudly "implemented AI." The vendor decks were impressive. The pilot is successful. The board is satisfied.

But here Ravi was, watching a loan approval—already assessed by AI in **three seconds**—wait **fourteen days** for human signatures.

Seven signatures. Four departments. Endless forwarding of emails that said nothing more than, *"Looks fine to me."*

That's when Ravi realized the uncomfortable truth:

The AI wasn't slow. The organization was.

They hadn't redesigned the system. They had **bolted AI onto a 20th-century machine** and expected magic.

The Hidden Cost No One Was Measuring

Ravi called it *Efficiency Debt.*

Every legacy rule, every silo, every approval layer had once been created to manage human limitations—slow analysis, inconsistent judgment, missing information.

But now those same safeguards had become friction.

The AI could see patterns across millions of data points. Humans were still trapped inside org charts drawn decades ago.

So, Ravi did something radical.

Instead of asking, "Where else can we use AI?"

He asked,

"If we were starting this company today—with AI as a given—would this process exist at all?"

The Audit That Changed Everything

The first two weeks were uncomfortable.

Leaders from finance, operations, HR, and IT gathered in a room—not to talk about tools, but to **map how work actually moved**.

What they saw was sobering.

Decisions crawled through linear paths designed for paper files. Data lived in departmental graveyards, not shared intelligence. People were hired for volume, not judgment. Customers were heard only after they complained.

AI had entered the building—but it had been treated like a guest.

Ravi made a call that stunned the room:

"AI isn't visiting anymore. It lives here now. And we're redesigning the house around it."

Zero-Based Redesign: Burning the Old Blueprints

They stopped optimizing. They started **rebuilding from zero**.

Approval workflows were the first to fall.

Ten days became one. Four layers became one human-in-the-loop. AI summarized everything. Flagged risks. Highlighted anomalies.

Humans no longer review *everything*. They reviewed **exceptions**.

Authority moved closer to the edges. Verification shifted from double-checking people to auditing algorithms. Incentives changed leaders were now rewarded for **decision velocity**, not team size.

Something unexpected happened.

People felt lighter. Smarter. More responsible—not less.

The Slow, Deliberate Transformation

Ravi refused to rush it.

The first three months were a sandbox. Low-risk areas. Safe experimentation. The real metric wasn't productivity—it was **psychological safety**.

Were people curious... or afraid?

Next came the core systems—sales, supply chain, forecasting. Old and new ran side by side. Comparisons were brutal but honest.

By month ten, there was no going back.

Legacy job descriptions were retired. Performance reviews have changed. A new phrase entered the organization's language:

Human–AI Partnership.

When the System Started Thinking Like a Brain

Data silos disappeared.

Marketing signals flowed into supply decisions in real time. Customer behavior reshaped inventory before shortages happened. The organization began to resemble a **neural network**, not a hierarchy.

Hiring changed, too.

Years of experience mattered less than one question:

"Can this person think clearly and instruct AI well?"

Top performers didn't work harder. They became architects.

Customers noticed the difference.

Most interactions were instant and automated. But when empathy was needed—real human empathy—it was there.

Faster service. Deeper loyalty.

Measuring What Actually Matters Now

Ravi stopped tracking hours worked.

Instead, he watched:

- How fast insight turned into action
- How much time people spent thinking versus searching
- How fresh information was when decisions were made
- How many roles had evolved from tasks to outcomes

The numbers told a new story.

But the real change was visible in meetings.

People questioned more. Deferred less. Owned decisions fully.

The Question Every Leader Must Answer

One evening, long after most had left, Ravi stood alone in the office.

On one wall hung the old organizational blueprint—dated 1987.

On the other, a living system designed for 2026.

AI hadn't transformed the company.

Leadership had.

By choosing to redesign the system—not the tool.

The question, Ravi realized, was no longer about technology.

It was this:

Will you keep managing the future with blueprints built for human limitation— or will you redesign for intelligence that now lives among us?

The next move, as always, belonged to the leader.

Top 5 Key Learnings

- AI rarely slows organizations; the real delay comes from **legacy approvals, policies, and decision structures** built for a pre-AI world. Leaders must remove this human latency by simplifying decision flows.
- AI fails when it is treated as **an add-on to old processes**. Real value comes when organizations redesign work, assuming AI is part of the system from the beginning.
- As AI handles data analysis and pattern detection, the **human role shifts from reviewing outputs to reviewing thinking**, assumptions, and trade-offs, while remaining accountable for decisions.
- The success of AI depends heavily on **organizational culture**. Psychological safety, experimentation, and learning-oriented environments allow teams to adopt AI effectively.
- Transformation becomes visible when leaders **change what they measure**, focusing on faster decisions, saved cognitive effort, and roles evolving from execution toward judgment.

Here are the **Top 5-chapter key takeaways:**

- *AI introduces a third thinking system that leaders must learn to direct.*

  Leadership now requires fluency across intuition, deep reasoning, and AI intelligence—without surrendering human judgment, ethics, or accountability.

- *Speed without sensemaking creates risk, lock-in, and false progress.*

  Moving fast with AI feels powerful, but without meaning, reflection, and system redesign, it leads to dependency, anxiety, and long-term rigidity.

- *AI success depends more on human capabilities than on technology.*

  Sensemaking, emotional resilience, learning velocity, and trust—not models or platforms—determine whether AI scales or silently fails.

- *Brilliant leaders must evolve identity, not just skills.*

  The AI age rewards leaders who shift from "having answers" to exercising judgment, asking better questions, and leading through uncertainty.

- *Organizations must redesign systems, metrics, and culture—not just add AI.*

  When workflows, governance, incentives, and measurements are rebuilt around human–AI partnership, AI moves from impressive demos to lasting advantage.

Part II

S — Making Meaning in Complexity

Sensemaking Intelligence (Self + System Awareness)

Introduction

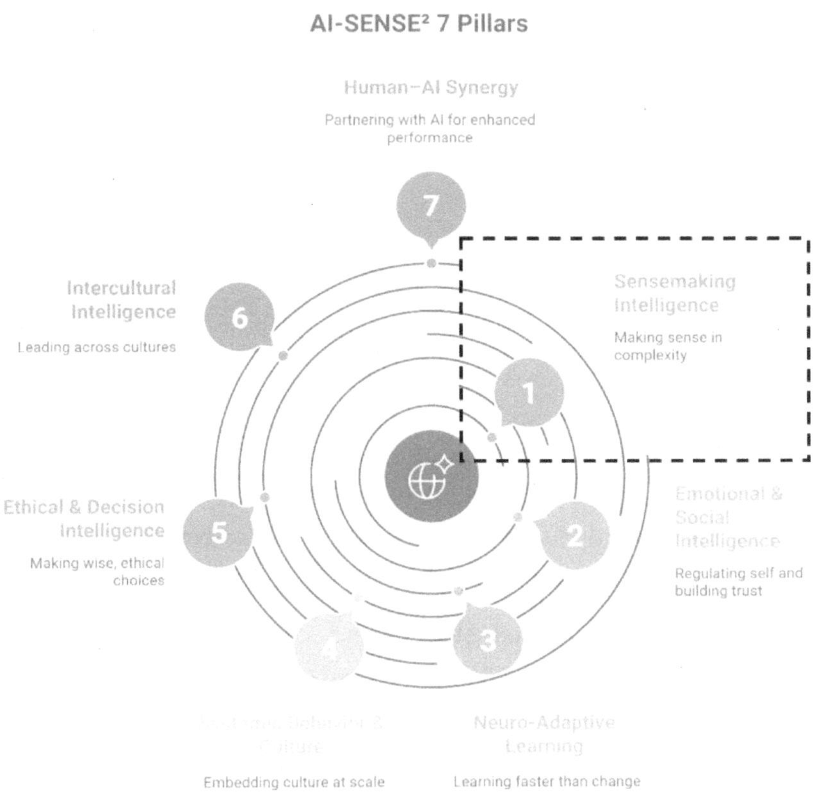

The greatest risk leaders face in the AI era is not a lack of information.

It is the **illusion of clarity**.

Dashboards multiply. Metrics refresh in real time. Predictions arrive before questions are fully formed. Yet despite this abundance of data, many leaders feel less certain than ever. Decisions become reactive. Context disappears. Action begins to replace understanding.

In a world saturated with signals, **meaning quietly fades**.

Part II—**SENSE** invites leadership to slow down—not to resist AI, but to **see reality more clearly**.

This section introduces **Sensemaking Intelligence** as a foundational leadership capability in the algorithmic age. It reframes awareness as more than observation and insight, as more than analysis. Sensemaking is the disciplined ability to interpret what is truly happening—across self, system, and consequence—before deciding what action is required.

Artificial intelligence can surface patterns.

It can detect anomalies.

It can even recommend actions.

But AI cannot determine what those patterns **mean**, why they **matter**, or how they will unfold within human systems. That responsibility remains deeply human.

Across these chapters, readers explore why greater volumes of data often produce **less clarity rather than more**. Leaders learn how to move beyond surface-level metrics to understand the deeper signals shaping organizations and decisions. They discover the difference between merely seeing information and truly **sensing what is unfolding beneath it**.

The section also expands awareness inward. Leaders examine how their own assumptions, emotional states, and cognitive biases influence interpretation—sometimes more powerfully than the data itself. True sensemaking begins with self-awareness and then extends outward to the broader system.

Through curiosity, reflection, and disciplined interpretation, leaders begin to bridge the gap between machine intelligence and human judgment.

By the end of Part II, leaders no longer feel compelled to chase every metric or react to every alert. Instead, they develop the capacity to **pause, frame complexity, and interpret reality with clarity**—using AI as a partner in understanding rather than a driver of constant reaction.

Leaders stop drowning in dashboards and begin to **interpret the world around them—calmly, thoughtfully, and with intention**.

Story: 6

Sense Before You Decide: Leadership Awareness in the Algorithmic Age

> **"Between stimulus and response, there is a space. In that space is our power to choose our response."**
>
> **- Viktor Frankl:**

Imagine waking up and scrolling through your phone.

A headline catches your eye—it perfectly confirms what you were already thinking. You feel that surge of satisfaction, almost like the world is nodding along with you.

Later, a dramatic news story makes your heart race.

You feel alarmed, anxious, maybe even outraged. By evening, you've spent hours in a whirlwind of notifications, your mind buzzing with thoughts that may not even be your own.

This isn't just a random experience; it's the invisible architecture of an algorithmic world.

Algorithms, designed to predict patterns and capture attention, have become the new landscape in which leaders and teams operate.

They know your preferences, your emotional triggers, and even your blind spots. They curate the world to match the predictable parts of your mind—but in doing so, they amplify your biases, your blind spots, and sometimes, your worst instincts.

Yet, there is something the algorithm cannot predict: conscious awareness.

Your ability to pause, notice, and choose your response.

This is the essence of **AI Agility**: mastering self-awareness while leveraging AI as a partner to make smarter, ethical, and more creative decisions.

How a Leader Applies It: Satya Nadella's Example

Consider **Satya Nadella, CEO of Microsoft**. He transformed Microsoft's culture by combining self-awareness, empathy, and AI-driven insight. Nadella recognized that traditional decision-making was too slow for a fast-changing tech world. He encouraged leaders to:

- **Reflect Before Reacting:** Leaders pause and analyze data rather than respond impulsively.
- **Leverage AI for Insight:** Microsoft uses AI to analyze vast datasets, surface trends, and support strategic decisions, but the final judgment always rests with the human leader.
- **Build Empathy and Awareness:** Nadella emphasized emotional intelligence, ensuring decisions consider people, ethics, and impact—not just efficiency.

How Teams Can Apply AI Agility

Self-awareness is often talked about as a leadership skill, but its real power shows up when it becomes a shared team habit. Teams that work with AI effectively don't just rely on faster data or smarter tools—they develop collective awareness about *how* they think, decide, and react under pressure.

Before locking in important decisions, these teams pause together. They take a moment to reflect on assumptions, emotional reactions, and hidden biases—both human and algorithmic. AI-generated insights are not accepted at face value; they are discussed, challenged, and placed in context. This shared reflection slows impulsive reactions without slowing progress.

Over time, teams begin to naturally distribute thinking roles. While AI might surface patterns or probabilities, different team members consciously bring complementary perspectives. One person grounds the discussion in facts, another raises ethical or long-term implications, and another tunes into customer impact or team sentiment. Decisions stop coming from a single viewpoint and start emerging from a richer, multi-lens conversation.

This approach also helps teams uncover blind spots. Instead of repeating familiar patterns, they deliberately invite perspectives that are missing—different functions, cultures, or experiences. AI may highlight trends, but humans ensure those trends are interpreted with empathy and situational understanding.

Emotion plays a role, too. In fast, digital environments, stress and overload quietly shape thinking. High-performing teams make space for brief emotional check-ins, acknowledging when pressure or fatigue is affecting judgment. This awareness prevents reactive decisions and strengthens trust within the group.

When human awareness and AI insight work together, teams move with greater clarity and confidence. They act faster without becoming careless, and smarter without losing their values. AI no longer replaces judgment—it supports it, freeing people to focus on what matters most: thoughtful strategy, creative problem-solving, and genuine human connection.

In essence: AI Agility is where emotion meets strategy, curiosity fuels growth, and human intelligence works in harmony with AI.

Teams and leaders who master this triad—self-awareness, collective reflection, and AI-powered insight—gain a decisive advantage in the modern, algorithmic world.

Here are the **top 3 key takeaways**:

- **Awareness must come before decisions in an algorithmic world.**

  Algorithms amplify bias and emotion, but leaders and teams retain a unique advantage: the ability to pause, notice triggers, and choose consciously before acting.

- **AI should inform judgment, not replace it.**

  As shown by **Satya Nadella**, the most effective leaders use AI for insight and pattern recognition while keeping human responsibility for empathy, ethics, and final decisions.

- **AI Agility is a team capability, not just a leadership trait.**

  Teams that combine self-awareness, collective reflection, and AI-powered insight make smarter, faster, and more ethical decisions—turning AI into a thinking partner rather than a decision-maker.

Story: 7

The Signal in the Noise: Why Leaders with More Data Often Understand Less

> **"If you don't know what to change, you don't know what to manage."**
>
> **- Eli Goldratt:**

The Signal in the Noise

In March 2024, **Ananya Iyer** had a problem most leaders quietly recognize—but rarely articulate.

She was drowning in data.

As **Chief Strategy Officer at Prakriti Biosciences**, a fast-growing biopharma company, Ananya had access to everything modern leadership promised:

AI-powered dashboards for real-time market intelligence, clinical, regulatory, sales, supply-chain, and sentiment analytics

And yet... she understood **less than ever**.

Sales were up. The doctor's confidence was slipping. Clinical trials looked strong. Patient feedback was turning uneasy. Social media conversations felt "off."

Everything was measurable. Nothing was clear.

"We have more data than we've ever had," Ananya said in one leadership meeting, "So why do I feel more blind than before?"

No one had an answer.

When AI Shows You Everything—but Explains Nothing

Two weeks later, the crisis hit.

Prakriti's flagship autoimmune drug, **Immunexa**, was linked to a cluster of severe adverse reactions. The drug regulator initiated a review. Investor confidence dropped overnight.

Looking back, the signals had been there all along:

- Patients discussing "unusual reactions" in regional health forums
- Doctors calling the medical helpline about unexpected drug combinations
- Insurance partners are slowing approvals
- Prescribing patterns subtly shifting across Tier-2 cities

Each signal alone looked insignificant. Together, they told a story.

The AI systems had flagged every data point. The dashboards had displayed every metric.

But **no one had made sense of what it meant**.

"We weren't lacking data," Ananya told the emergency leadership meeting. "We were lacking sensemaking."

The Shift: From Analysis to Sensemaking

Ananya reached out to **Dr. Aarav Menon**, an organizational psychologist known for working with Indian enterprises navigating high-complexity environments.

Their first conversation didn't start with dashboards. It started with a question.

"When everything contradicts," Aarav asked, "what question are you actually trying to answer?"

That reframed everything.

He explained the difference:

Analysis asks: *What patterns exist in the data?*

Sensemaking asks: *What is changing in the system—and which signals matter?*

From that day, leadership meetings changed.

They stopped starting with charts. They started with questions:

- What would we notice if something important were shifting?
- What surprised us this week?
- What doesn't quite fit our assumptions?

The panic softened. Curiosity returned.

Why Smart Teams Still Miss the Big Picture

Ananya soon discovered another truth:

Smart individuals see fragments of reality. Only **collective sensemaking** reveals the whole picture.

At Prakriti, leaders from R&D, Sales, Medical Affairs, and Compliance were all "right"—but only partially.

So, they redesigned leadership discussions:

- Silent reflection before debate
- Sharing signals instead of conclusions
- Pattern-hunting across functions
- Hypothesis-building before decision-making

Disagreements didn't disappear. They became productive.

People stopped defending silos. They started assembling meaning.

The Layer Most Leaders Miss: Power & Narratives

Even with clarity, resistance emerged.

Finance wanted aggressive growth. Sales wanted more time in difficult markets. R&D wanted science to lead strategy.

Each group had data. Each had influence. Each had a narrative.

Ananya learned the hardest lesson of sensemaking:

Truth alone doesn't win. Narratives do.

Instead of pushing strategy, she invited **co-sensemaking**:

- "What are you seeing that we might be missing?"
- "What outcome do we all want to protect?"
- "Where are our assumptions clashing?"

Strategy stopped feeling imposed. It started feeling shared.

Redefining AI's Role

The final breakthrough came when Prakriti reframed AI's purpose.

AI is exceptional at: • Detecting patterns at scale • Flagging anomalies early • Modelling complex scenarios

AI struggles with: • Context • Ethics • Meaning • Human consequences

So, they redesigned their operating model:

AI surfaces signals. Humans interpret meaning. Teams make judgment calls.

That single shift changed everything.

Eighteen Months Later

- No surprise crises
- Faster, clearer strategic decisions

- Higher trust across leadership
- Improved margins with focused growth
- Teams asking better questions—not chasing more dashboards

Ananya reflected during a company town hall:

"We didn't need more AI. We needed better sensemaking."

The Leadership Insight

In an AI-driven world:

• Data is abundant • Information is cheap • Understanding is rare

The leaders who thrive won't be the most technical.

They'll be the ones who can:

- Ask sharper questions
- Distinguish signal from noise
- Integrate diverse perspectives
- Navigate power and narratives
- Partner with AI without surrendering judgment

That capability has a name:

Sensemaking Intelligence

And in today's world, it's not optional.

It's leadership survival.

Here are the **top 3 key takeaways**:

- **More data does not create more understanding—sensemaking does.**

  Leaders often miss critical shifts not because signals are absent, but because no one is interpreting how those signals connect and what they mean together.

- **AI surfaces signals, but humans must create meaning.**

  AI excels at pattern detection and anomaly spotting, but context, ethics, and judgment remain human responsibilities. Clarity emerges only when humans stay in the loop.

- **Collective sensemaking beats individual intelligence.**

  Real insight comes when leaders integrate perspectives across functions, surface assumptions, and navigate power and narratives together—turning noise into shared understanding.

Story: 8
When Seeing Isn't Enough: The Rise of Sensemaking Intelligence

> **"The more data we collect, the less information we have."**
>
> **– Russell Ackoff**

The city of **Aethel** had always been alive.

Its streets pulsed with movement—vendors calling out prices, commuters rushing through transit hubs, children weaving through public squares. From above, it looked vibrant and safe. From inside the city's security command centre, it felt very different.

Rows of operators sat before walls of screens. Hundreds of live CCTV feeds flickered at once—parks, markets, stations, municipal buildings. Every screen demanded attention. Every moment felt urgent.

What looked like vigilance was, in truth, paralysis.

Despite more guards, longer shifts, and added patrols, incidents kept slipping through. A pickpocket vanished into a crowd. An abandoned bag sat unnoticed until panic spread. Someone crossed into a restricted zone, seen only after the damage was done.

The problem wasn't effort. It wasn't technology. It was **sensemaking**.

Sensemaking Intelligence — Making Meaning in Complexity

Human attention was drowning in data. Operators weren't failing because they lacked skill, but because the system forced them to treat **everything as equally important**.

The city council realized that incremental fixes would never solve a systemic problem. They didn't need more eyes. They needed **better meaning-making**.

That realization marked Aethel's first shift—toward what would later be called **AI-SENSE²**.

Choosing Partnership Over Replacement

Instead of asking how AI could replace humans, the city asked a different question:

How can AI help humans see what matters most?

They recalibrated every algorithm with clear intent. Cameras and models were aligned to detect only what truly mattered—intrusions, unattended objects, and abnormal patterns. Noise was filtered. Signals were sharpened.

This wasn't surveillance for control. It was an **ethical decision, intelligence** in action.

Ethical & Decision Intelligence — Wise, Context-Aware Choices

From day one, ethical boundaries were explicit. Privacy was protected. Transparency was enforced. The AI existed to safeguard citizens—not to watch them indiscriminately.

Trust grew quietly but steadily.

From Monitoring to Meaning

As the system went live, something unexpected happened.

The AI could scan dozens of feeds at once—something no human could do—but it didn't act alone. Instead, it **surfaced insights**, patterns, and probabilities.

Humans reclaimed their role—not as watchers, but as **interpreters**.

Emotional & Social Intelligence — Regulating Self, Building Trust

With cognitive overload reduced, operators felt calmer. Alert fatigue faded. Stress dropped. They could think again.

When a warning appeared, it mattered.

Instead of reacting emotionally to constant noise, teams responded with composure. Psychological safety emerged—not because the job got easier, but because it became **humanly sustainable**.

Learning Faster Than the City Changed

Not every alert was perfect.

Some were false positives. Some responses were slow. But none were ignored.

Every incident became feedback.

Neuro-Adaptive Learning — Learning Faster Than Change

The AI improved with data. The humans improved with reflection.

Operators learned how the system "thought." They understood what it weighed heavily—and what it missed. Over time, human judgment sharpened. AI precision increased.

Learning velocity became the real advantage.

From Reaction to Anticipation

With clarity restored, something profound shifted.

Operators stopped chasing incidents. They began **anticipating them**.

Systemic Behavior & Culture — Embedding Change at Scale

New habits formed:

- Proactive scanning instead of passive watching
- Calm coordination instead of rushed escalation
- Clear ownership instead of confusion

These behaviors weren't enforced. They **emerged**—reinforced by a system designed to support human cognition, not overwhelm it.

Culture changed not through mandates, but through **better-designed work**.

Humans and Machines, One Team

In the command center, boundaries softened.

Operators, field officers, and AI worked as one synchronized system.

Intercultural × Human–AI Intelligence — Leading Across Humans and Machines

The AI flagged anomalies. Humans assessed context. Together, they acted.

Decisions were faster—but also wiser. Interventions were precise, avoiding unnecessary intrusion while ensuring safety.

Accountability remained human. Capability became shared.

Flow, Focus, and Sustainable Performance

Over time, results spoke quietly but clearly.

Response times improved. Crime dropped in monitored zones. Operator burnout declined.

People entered states of deep focus—what athletes call **flow**. Not because the work disappeared, but because attention was finally protected.

AI carried the cognitive weight. Humans carried the responsibility.

The City That Learned to Lead with AI

Aethel did not become safer because it added more cameras or more people.

It became safer because it learned to **align human leadership, neuroscience, and AI collaboration**.

AI-SENSE² was never about technology.

It was about:

- Sensemaking over surveillance
- Ethics over efficiency
- Learning over control
- Partnership over replacement

The city didn't just learn to see better.

It learned to **lead smarter**.

Here are the **top 3 key takeaways**:

- **Seeing more data is useless without sensemaking.**

  The real failure wasn't a lack of technology or effort—it was treating all signals as equal. Leaders must design systems that help humans understand what truly matters.

- **AI creates value only when it augments human judgment, not replaces it.**

  AI is powerful at filtering noise and surfacing patterns, but meaning, ethics, and accountability must remain human-led to make decisions wise, not just fast.

- **Sustainable performance comes from human–AI partnership.**
- When AI reduces cognitive overload and humans focus on interpretation, learning, and ethics, teams shift from reaction to anticipation—and performance improves without burnout.

Story: 9

Leading with Curiosity: How AI Strengthens Human Sensemaking and Judgment

> **"The important thing is not to stop questioning. Curiosity has its own reason for existing."**
>
> **– Albert Einstein**

Curiosity is not a soft trait or a passing mood. It is the engine that drives learning, innovation, and meaningful progress. Every breakthrough begins with a deeper question—*Why does this work the way it does? How could it be different? What if the rules changed?* When these questions are asked consistently, curiosity turns into a disciplined way of thinking.

What makes this moment unique is that curiosity no longer has to be a solo effort. AI can step in as a quiet but powerful partner—a Curiosity Co-Pilot that helps you go beyond obvious answers and explore paths you may never have noticed on your own.

Imagine being stuck on the surface of an idea. You understand the topic, but nothing new is emerging. This is where AI becomes useful. By sharing a simple theme or problem, AI can immediately respond with layers of *why*, *how*, and *what-if* questions that cut beneath the surface. Each question opens a new direction, turning a flat topic into a landscape of possibilities. Curiosity grows not because answers are given, but because better questions appear.

As you go deeper, another challenge emerges—hidden assumptions. We all carry beliefs we rarely question, and these blind spots quietly limit our thinking. AI can act as a disciplined challenger, calmly pointing out what you may be taking for granted. When an idea is tested this way, curiosity shifts from comfort to rigor. You are no longer just exploring—you are refining how you think.

Over time, AI can take on the role of a structured questioning partner, much like a modern Socratic guide. It can help you analyze causes, evaluate evidence, reflect on how your views have changed, and imagine creative alternatives. This turn learning into a dialogue instead of a one-way process. Curiosity expands because your thinking moves across multiple dimensions rather than staying in a single lane.

Another powerful effect comes from exposure to contrast. Curiosity weakens when we only see one perspective. AI can instantly surface opposing viewpoints, cultural interpretations, or alternative arguments. When you encounter a difference without friction, your mind naturally leans in. Nuance replaces certainty, and curiosity becomes more resilient and open-ended.

Some of the most exciting moments of curiosity happen when unrelated ideas collide. AI is especially good at forcing these intersections—bringing together concepts that would rarely meet on their own. When empathy meets logistics, or biology meets architecture, new questions emerge. These unexpected connections train the mind to explore beyond familiar boundaries, where innovation often lives.

Curiosity also deepens when ideas are tested in realistic situations. By simulating real-world scenarios—ethical dilemmas, leadership challenges, or policy decisions—AI allows you to explore consequences without real-world risk. Each scenario invites deeper questioning, sharpening judgment and perspective through experience rather than theory.

Over time, curiosity becomes strongest when it is practiced daily. Small habits—such as building a personal question bank or starting each day with a set of reflective and creative prompts—slowly rewire how the mind approaches information. What once felt passive becomes an active inquiry.

At its best, AI becomes more than a tool. It becomes a coach—one that challenges your thinking, pushes back on easy conclusions, and invites you to go further. Used this way, AI does not replace curiosity. It strengthens it, stretches it, and keeps it alive.

In a world flooded with answers, curiosity is the true advantage. With AI as a co-pilot, questioning becomes deeper, thinking becomes sharper, and exploration becomes limitless.

What idea are you ready to challenge with your new AI Curiosity Partner today?

Here are the **key takeaways**:

- **Curiosity is a leadership skill—and AI can actively strengthen it.**

  By generating deeper *why, how, and what-if* questions, AI helps leaders move beyond surface thinking and explore ideas with greater depth and rigor.

- **AI sharpens judgment by challenging assumptions, not just providing answers.**

  Acting as a devil's advocate or Socratic partner, AI exposes blind spots, presents opposing views, and forces clearer reasoning, making thinking more disciplined and honest.

- **Breakthrough thinking emerges when humans direct AI with intention.**

  When used deliberately—as a curiosity coach, scenario simulator, and idea connector—AI expands creative exploration while humans remain responsible for meaning, ethics, and decisions.

Story: 10

What Leaders Can Learn from an AI-Driven Fusion System: A Lesson in Decision-Making Under Complexity

> **"The whole is greater than the sum of its parts."**
>
> **– Aristotle**

In today's world, the real challenge is no longer a lack of data. It is the opposite. We are surrounded by dashboards, alerts, reports, models, and constant streams of information. Every system promises better visibility. Every tool promises sharper insight. Yet many leaders feel more uncertain than ever. Decisions slow down. Teams debate longer. Clarity becomes harder to find.

A powerful lesson comes from a fictional maritime defence force operating across a vast and sensitive ocean region. The setting may be imagined, but the challenge is familiar. This force relied on an expanding network of sensors—crewed ships, autonomous vessels, surface radars, underwater systems, and electronic surveillance tools. Each sensor was valuable. Each delivered important signals. But together, they also created confusion.

Different sensors produced different interpretations. Some were more accurate in calm waters. Others performed better in harsh weather. Some detected patterns that others missed. At times, the same situation triggered conflicting alerts. Was it a natural object? A vessel? A threat? The more signals they received, the less certain they felt.

This mirrors what many organizations experience today. Multiple analytics reports point in slightly different directions. Experts disagree. AI models trained on different datasets produce different predictions. More data does not automatically create better decisions. Without structure, it simply amplifies noise.

The turning point came when the force stopped searching for a single "perfect" model. Instead, they redesigned their approach around intelligent fusion. Their goal was simple: turn complexity into clarity.

First, they allowed multiple AI models to analyze the same incoming data. Each model produced its own confidence score based on its strengths. Instead of averaging those scores blindly, the system adjusted their influence dynamically. If a model had historically performed well in similar environmental conditions, its voice carried more weight. If another model struggled in certain weather patterns, its influence was reduced in those moments. Domain knowledge shaped which signals mattered most. No model was discarded. Each contributed where it was strongest. The lesson was clear: strength does not come from eliminating diverse views, but from integrating them wisely.

Next, they examined the sensors themselves. Not all data sources are equal. Some sensors were newer and more precise. Others were older or limited by environmental conditions. The AI began evaluating the quality of each data stream—its resolution, reliability, and relevance to the current situation. Higher-quality inputs carried greater weight. Weaker signals were not ignored, but they were treated carefully. This created a unified confidence score built on quality, not just volume. The insight for business is simple: treating all data as equal weakens decisions. Respect data quality, not just data quantity.

The final layer was context. Human experts naturally consider history and environment when making decisions. The AI learned to do the same. It incorporated known operational patterns, past mission data, seasonal behaviors, and environmental conditions. A signal that looked threatening in isolation might appear routine when seen in historical context. Or the opposite might be true. Context transformed raw signals into meaningful interpretation.

This approach changed everything. Previously, a system might report extremely high certainty that an object was non-natural, strong likelihood of hostile intent, and moderate confidence in classification. Each metric sounded impressive on its own. But without reconciliation, confidence could be misleading. Through fusion—balancing model disagreement, sensor reliability, and situational context—the final assessment became grounded and earned rather than assumed.

The greatest advantage was speed. By intelligently layering these fusion processes, the force reduced the time required to resolve contradictions and reach situational understanding. In high-pressure environments, speed combined with accuracy becomes strategic power. The ability to move quickly without sacrificing judgment is what separates reactive systems from resilient ones.

Yet one principle remained unchanged. The AI did not replace human judgment. It supported it. Leaders received clearer insights, transparent confidence levels, and far less cognitive overload. They could see not just what the system concluded, but why. The final decision still rested with people—but now it was informed by synthesized intelligence rather than fragmented signals.

The deeper lesson extends far beyond maritime defense. In a world overflowing with information, clarity does not come from finding one perfect answer. It comes from intelligently fusing perspectives. The future belongs to leaders and teams who can integrate diverse signals, weigh them thoughtfully, and act with confidence grounded in context.

Whether you are leading:

- An AI initiative
- A product organization
- A transformation program
- Or a complex team under uncertainty

The principle holds:

Integrate strengths. Apply context. Decide with confidence.

That is how intelligent systems—and intelligent leaders—win in complexity.

These questions are designed to **provoke reflection, shift mindset, and drive behavior change**—not just insight.

1. Sensemaking Intelligence

Making Meaning in Complexity

1. *When faced with conflicting signals—from people, data, or AI—what meaning are you currently making, and what other interpretations might be equally valid?*

Purpose: Expands perspective and reduces premature certainty.

2. Emotional & Social Intelligence

Regulating Self, Reading Others, Building Trust

2. *What emotional reaction does uncertainty trigger in you, and how does that reaction shape the way your team experiences your leadership?*

Purpose: Builds emotional awareness and psychological safety.

3. Neuro-Adaptive Learning

Learning Faster Than Change

3. *Which belief, habit, or way of thinking helped you succeed in the past—but may now be slowing your ability to adapt?*

Purpose: Encourages unlearning and cognitive flexibility.

4. Sensemaking × Emotion

Integrating Thought and Feeling

4. *How do your emotions influence the stories you tell yourself about what the data—or the AI—is saying?*

Purpose: Connects emotional intelligence with sensemaking.

5. Systemic Behavior & Culture

Embedding Change and Making It Stick

5. *If someone observed your team for one week, what daily behaviors would signal that transformation is truly happening—or not?*

Purpose: Shifts focus from intention to observable behavior.

6. Ethical & Decision Intelligence

Wise, Context-Aware Choices

6. *Even if an AI recommendation is highly confident, what values or long-term consequences must you still weigh before acting?*

Purpose: Reinforces ethical judgment over blind automation.

7. Human–AI Partnership

Leading with Machines, Not Against Them

7. *Which decisions should never be delegated to AI—and why are those decisions central to your role as a human leader?*

Purpose: Clarifies boundaries between human responsibility and AI augmentation.

8. Intercultural Intelligence

Leading Across Cultures and Contexts

8. *How might this decision be interpreted differently across cultures, roles, or regions—and whose perspective is missing right now?*

Purpose: Reduces bias and increases inclusion at scale.

9. Systemic Integration

From Insight to Enterprise Impact

9. *What systems, incentives, or rituals must change so that this way of thinking becomes the norm rather than an exception?*

Purpose: Anchors leadership insight into organizational systems.

Why These Questions Matter

These are not diagnostic questions. They are **transformational questions**.

They help leaders:

- Move from **data to meaning**
- From **reaction to regulation**

- From **tools to trust**
- From **change initiatives to cultural adoption**
- From **AI usage to human-AI wisdom**

Here are the **top 3 key takeaways**:

- **More data does not improve decisions—fusion does.**

  Clarity emerges when diverse signals, models, and perspectives are intelligently integrated with context, not when leaders chase a single "best" answer.

- **AI creates an advantage only when it supports human judgment.**

  The most effective systems reduce cognitive overload, surface meaningful insight, and leave ethics, accountability, and final decisions firmly with humans.

- **Leadership in complexity is about sensemaking, not certainty.**

  Leaders who fuse viewpoints, regulate emotions, apply context, and align systems can act faster and wiser—turning complexity into a strategic edge.

Here are the **top 5 key takeaways**:

- **Awareness must come before action.**

  In an algorithmic world, speed without self-awareness leads to reactive decisions. Leaders must pause, notice biases and emotions, and choose consciously before acting.

- **More data does not create clarity—sensemaking does.**

  Dashboards and signals only become valuable when humans interpret how they connect, what has changed, and why it matters.

- **AI should augment judgment, not replace it.**

  AI excels at pattern detection and learning, but meaning, ethics, and accountability must remain human responsibilities.

- **Sensemaking is a collective capability.**

  The best decisions emerge when teams fuse diverse perspectives, surface assumptions, and interpret signals together—no single leader or model sees the whole picture.

- **Curiosity and continuous learning are leadership disciplines.**

  Leaders who ask better questions, learn in small cycles, and integrate context consistently turn complexity into a strategic advantage.

In essence:

Sense early, think together, learn continuously, and decide with human judgment at the center.

Part III

E — Emotional & Social Intelligence

Regulating Self, Leading Humans Through AI Change

Introduction

AI-SENSE² 7 Pillars

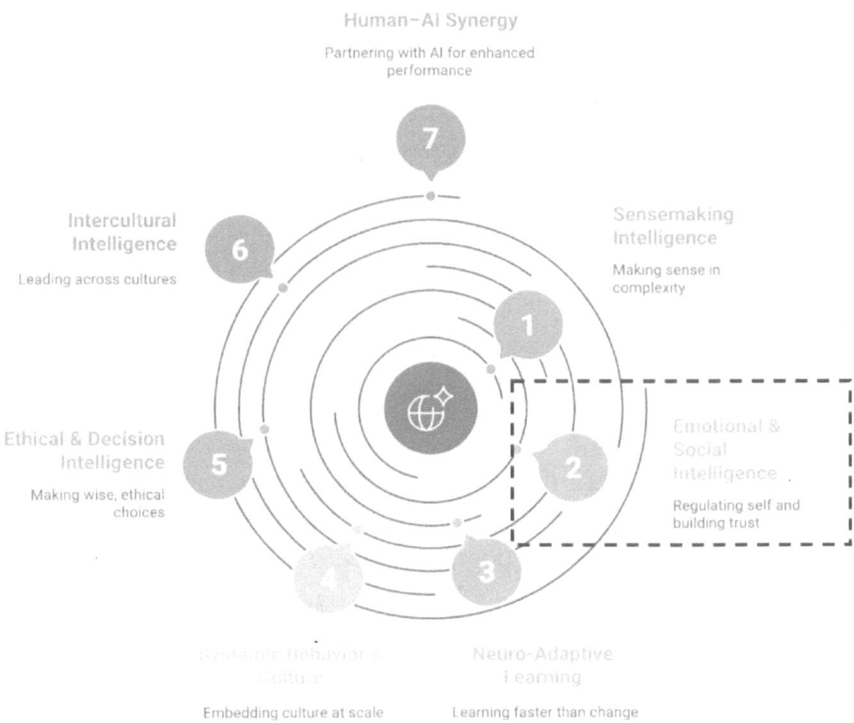

AI transformations rarely fail because the technology is wrong.

More often, they fail because the **human system surrounding the technology becomes overwhelmed**.

Fear appears before strategy.

Identity feels threatened before processes change.

Trust begins to erode quietly while leaders remain focused on models, tools, and implementation roadmaps.

Part III—**Emotional & Social Intelligence** brings leadership back to the human center of AI transformation.

This section addresses a truth many organizations overlook: adopting AI is not primarily a technical or rational exercise. It is, at its core, an **emotional transition**.

Before people evaluate the usefulness of AI, they often experience anxiety about relevance, control, and purpose. Before teams align around new ways of working, they look to leaders for signals of safety, clarity, and reassurance.

Across these chapters, readers explore why **emotional intelligence must precede artificial intelligence** in successful transformations. Leaders learn how unmanaged stress, ego, and overconfidence can amplify automation bias and weaken decision quality. They also see why leadership presence becomes **more visible—not less—when AI enters decision systems**, and how even subtle behaviors influence whether AI is perceived as a tool for empowerment or a threat to human value.

The section also introduces one of the most dangerous patterns emerging in the AI era: treating AI like a *brilliant intern*—delegating thinking instead of retaining responsibility. These chapters guide leaders toward a healthier partnership, where AI accelerates insight while humans continue to provide judgment, ethics, and accountability.

Most importantly, this part shows that AI adoption rarely happens through mandates, dashboards, or performance metrics alone. It takes root through everyday leadership behavior—how leaders listen, explain decisions, handle uncertainty, and create space for learning.

When people feel safe, trusted, and included, intelligent systems begin to create meaningful impact.

By the end of Part III, leaders no longer view emotions as resistance that must be overcome. Instead, they recognize emotions as **signals that guide how leadership must respond during transformation**.

They understand that AI adoption is emotional before it is rational—and that **trust, identity, and psychological safety ultimately determine whether intelligence becomes real organizational impact**.

Story: 11

Why Emotional Intelligence Comes Before AI Intelligence

> **"Vulnerability is not weakness; it is our greatest measure of courage."**
>
> **– Brené Brown**

The Mirror Before the Mountain

Sarah Chen had built her career on answers.

For twenty years at Meridian Financial Services, she had learned what leadership looked like: confidence without hesitation, decisions made quickly, uncertainty handled with steady hands. She had survived market crashes, regulatory shocks, and endless restructurings. People trusted her because she *knew*.

So, when the CEO announced a company-wide AI transformation, Sarah did what she had always done. She smiled. She nodded. She told her team of forty-two analysts that everything would be fine.

That night, she didn't sleep.

At two in the morning, Sarah sat alone at her kitchen table, laptop glowing in the dark. Article after article scrolled past her tired eyes.

AI processes loan applications ten times faster than humans. Machine learning predicts market trends with unprecedented accuracy. Eighty percent of routine financial analysis could be automated by 2027.

A question tightened in her chest and refused to loosen.

If AI can do what my team does... what value do I bring?

She had spent decades mastering financial judgment—risk, nuance, mentorship. Now, algorithms were doing in seconds what her team took hours to complete. Worse, she didn't truly understand how they worked.

For the first time in her career, Sarah felt like an impostor.

The next morning, she stood before her team and delivered a polished presentation about "exciting opportunities ahead." Her voice was steady. Her slides were confident. Her hands trembled just enough to notice.

When Marcus, one of her most trusted analysts, asked, "How do you think this will change our roles?" she replied automatically, "We'll adapt. We always do."

The words felt empty.

That evening, her mentor, Dr. Patricia Wong, called for their regular check-in.

"How are you *really* doing with this AI transition?" Patricia asked.

Sarah tried to deflect. Then something cracked.

"I'm terrified," she admitted. "And I have no idea what I'm doing. But I can't let my team see that."

Silence lingered on the line.

"Why not?" Patricia asked gently.

"Because I'm the leader," Sarah said. "They need me to have answers."

"Do they?" Patricia replied. "Or do they need you to have the courage to figure it out with them?"

The question landed hard.

"You're trying to lead people through terrain you haven't walked yourself," Patricia continued. "You can pretend for a while. But eventually, everyone gets lost."

"So, what do I do?"

"You start with yourself. You coach *you* first."

That was the moment everything changed.

Over the next few weeks, Sarah began doing the work she had always avoided—looking inward. She started journaling privately, writing fears she had never allowed herself to name.

I'm afraid I'll become obsolete. I'm afraid of looking incompetent. I'm afraid my expertise no longer matters.

Seeing the words on paper didn't erase them—but it softened their grip.

With the help of an executive coach, Sarah realized the truth: her anxiety wasn't about AI. It was about identity. She had built her self-worth on being the expert. AI threatened that identity.

The work wasn't learning faster. It was learning differently.

She began practicing "productive discomfort." She attended AI workshops where she understood the least. She asked basic questions. She said, "I don't know."

It was humiliating. It was freeing.

She also noticed her triggers. Whenever uncertainty surfaced, she rushed to fix things—to decide quickly, to regain control. Her coach helped her pause and ask a different question:

Am I solving the problem—or my anxiety about not knowing?

The real shift happened during a team meeting.

Priya, a junior analyst, raised concerns about bias in a pilot AI credit model. Sarah felt the old reflex kick in—the urge to reassure, to resolve, to project certainty.

Instead, she took a breath.

"I don't know enough about these models either," she said. "And that's something we need to understand before moving forward."

The room went still.

Then she added, "Priya, would you lead an investigation? I'll make sure you get the support—and I'll learn alongside you."

After the meeting, Marcus stopped by her desk.

"That was different," he said. "When you stopped pretending, it gave us permission to stop pretending too."

Something loosened inside Sarah.

From that day on, she led differently.

AI initiatives became experiments, not mandates. Meetings became conversations, not performances. Mistakes became data, not failures.

She created spaces where questions were welcomed, and uncertainty was normal. She modelled learning publicly. She admitted when things didn't work.

And slowly, trust deepened.

Eight months later, her team wasn't the most technically advanced in the company—but it was the most adaptive. They learned faster because they were honest. They moved forward because they weren't afraid to pause.

When a peer later asked how she had led such a "smooth transition," Sarah laughed.

"It wasn't smooth," she said. "It was uncomfortable. But the moment I stopped pretending I had answers, everything changed."

"Doesn't that weaken your authority?" the peer asked.

"No," Sarah replied. "It changed it."

That night, Sarah updated her journal.

I used to think leadership meant knowing. Now I know it means learning. I used to think vulnerability was weakness. Now I know it's trust.

She closed her laptop and caught her reflection in the dark screen.

The mountain of AI transformation still stood ahead—complex, shifting, impossible to map fully.

But she no longer needed to pretend she had already climbed it.

She had faced the mirror first.

And that, she finally understood, was where leadership truly begins.

Here are the **3 key takeaways**:

- **Emotional intelligence must come before AI intelligence.**

  Leaders cannot guide others through AI-driven change if they haven't first acknowledged and regulated their own fear, identity threats, and uncertainty. Self-awareness is the foundation of credible leadership in the AI age.

- **AI threatens identity more than capability—and leaders must address that openly.**

  Resistance to AI is rarely about technology; it's about self-worth tied to being "the expert." When leaders separate identity from expertise, they unlock learning, adaptability, and trust for themselves and their teams.

- **Vulnerability transforms authority into trust.**

  By admitting "I don't know," inviting others to explore with her, and modeling learning publicly, Sarah created psychological safety. This allowed her team to adapt faster, think more honestly, and use AI responsibly—without fear or pretense.

Story: 12

The Brilliant Intern Trap: Why Leaders Must Learn to Think With AI, Not for It

> **"The true sign of intelligence is not knowledge but imagination."**
>
> **– Albert Einstein**

The Paradox of the "Brilliant Intern"

When Ravi integrated AI into his department's daily operations, the immediate results were intoxicating.

Dashboards turned green, response times plummeted, and the output volume tripled. It was as if every employee had been paired with a brilliant, tireless intern from the world's top university.

But three months later, Ravi noticed a disturbing trend. In strategy meetings, his senior leads began quoting the AI's "confidence scores" as if they were divine truths. They were outsourcing the ***logic*** of their roles, not just the ***labor***.

This is the central leadership challenge of 2026: **How do we use a tool that is often smarter than us without losing our own ability to think?**

1. The Shift: From Answer Machine to Thinking Partner

Most organizations treat AI as an **Oracle** (an entity that provides the "right" answer) or a **Utility** (like electricity—always on and invisible).

To remain Agile, leaders must reframe AI as a **Thinking Partner**. While a tool gives you the "What," a Coach pushes you on the "How" and the "Why."

Strategy: "Interrogate the Reasoning, Not the Result"

The trap is the "Recommendation." When AI suggests a market entry or a budget cut, it does so with linguistic perfection.

- **The AI Coaching Nudge:** Instead of asking, *"What is the best move?"*, a leader-coach trains the team to ask: *"What logic-path led to this suggestion?"*
- **Practical Application:** Use the "Three-Point Challenge." Ask the AI to provide:
- **Coaching Insight:** Judgment is a muscle. If you allow AI to lift the heavy weights of logic for you, your cognitive strength will atrophy.

2. Navigating the "Data vs. Wisdom" Divide

Ananya's team had a "perfect" data model for their 2026 expansion. The AI analyzed millions of data points and confirmed a go-ahead. But Ananya felt a tension in the room. The "terrain"—the cultural nuances and the team's burnout levels—wasn't in the data set.

Strategy: "Separate the Map from the Terrain"

AI provides the map (historical patterns, statistics, trends). The leader must walk the terrain (human emotions, ethics, and context).

- **The AI Coaching Nudge:** Ask the AI, *"What context is missing from your training data that might change a human's mind here?"*
- **Practical Application:** Before making a decision, list the "Invisible Factors"—empathy for the customer, internal team morale, and long-term brand legacy.
- **Coaching Insight:** Intelligence is about processing information. Wisdom is about knowing which information actually matters in the current moment.

3. Mastering the Probability Gap

Raj relied on an 80% success forecast for a product launch. When the project failed, he felt the AI had "lied." But the AI didn't lie; it simply reported that there was a 20% chance of failure—and that 20% became the reality.

Strategy: "Coach for Scenarios, Not Certainty"

Agility in 2026 isn't about being right; it's about being prepared to be wrong.

- **The AI Coaching Nudge:** Shift from *"What will happen?"* to *"What are the most likely failure modes for the 20% scenario?"*
- **Practical Application:** Use AI to build **Plan B and Plan C** simultaneously. Ask it to "Red Team" your primary strategy by assuming the AI's own prediction is wrong.
- **Coaching Insight:** Agility is not the accuracy of your prediction—it is the speed and grace of your recovery.

4. The Verification Ritual: Neutralizing Overconfidence

In a high-stakes boardroom, a team presented an AI-generated competitor analysis. It was brilliant, confident, and partially "hallucinated." The AI hadn't tried to deceive; it had simply filled a data gap with a plausible-sounding fiction.

Strategy: "Normalize Verification Over Trust"

In 2026, blind trust is a liability. Confidence in tone must be met with scepticism in process.

- **The AI Coaching Nudge:** Treat AI outputs as a "Draft 0" that requires a **Human-AI Verification Loop.**
- **Practical Application:** Implement a "Source-Check" ritual for every high-stakes output. If the AI cannot point to a verifiable data source or logic chain, it is treated as a creative suggestion, not a factual report.
- **Coaching Insight:** Confidence without verification is not intelligence—it is high-speed risk.

The Non-Delegable Core: What Must Remain Human?

As Ravi's team matured, they realized they were truly Agile only when they knew exactly where the AI stopped and where they began. There are some things the "Brilliant Intern" can never own:

1. **Ethical Responsibility:** AI calculates the "efficient" path; humans must choose the "just" path.
2. **Emotional Intelligence:** Only a human can feel the weight of a layoff or the spark of a true breakthrough.
3. **Accountability:** If the decision fails, you cannot blame the algorithm. You are the one who took the advice.

Final Reflection for the 2026 Leader

AI will only become more fluent, more convincing, and more integrated. The goal of the modern leader is not to compete with the machine's speed, but to protect the depth of human thought.

The best leaders of 2026 will not be those with the best prompts, but those who use AI as a whetstone to sharpen their own wisdom. The intern is here to stay—but the steering wheel must always remain in human hands.

The next right step is yours to take.

In an AI-saturated world, the scarcest skill is not information access—it is **meaning-making**.

Key thought:

Leaders who cannot interpret complexity will be overwhelmed by intelligence—human or artificial.

AI accelerates data. **Sensemaking turns data into direction.**

As AI takes over cognitive load, **emotional regulation and trust become the true productivity drivers**.

Key thought:

Teams don't fail because of poor tools; they fail because leaders cannot regulate fear, uncertainty, and ego in AI-driven change.

Psychological safety is not soft—it is **enterprise-critical**.

Experience decays faster than ever. What scales now is **neuro-adaptive learning**.

Key thought:

In the age of AI, the advantage belongs to leaders who can rewire themselves faster than the environment changes.

AI rewards curiosity, not certainty.

AI adoption without cultural design leads to **surface change and deep resistance**.

Key thought:

Strategy introduces AI. Culture decides whether it survives.

Sustainable transformation happens when **habits, rituals, and reinforcement systems** change—not just tools.

AI is not a tool to deploy—it is a **collaborator to design with**.

Key thought:

The future belongs to leaders who know what to delegate to machines—and what must remain human.

Judgment, accountability, and moral courage **do not scale through algorithms**.

Here are the **top 3 key takeaways:**

- **AI must be treated as a thinking partner, not an answer machine.**

  When leaders outsource judgment to AI's recommendations or confidence scores, their own reasoning muscles weaken. Real agility comes from interrogating *how* and *why* AI reaches conclusions—not blindly accepting *what* it suggests.

- **Data intelligence is not the same as wisdom—and leaders must bridge the gap.**

  AI provides maps (patterns, probabilities, forecasts), but humans must walk the terrain (context, ethics, emotions, culture). Decisions fail when leaders confuse statistical confidence with real-world understanding.

- **What remains human determines whether AI succeeds or fails.**

  Ethical responsibility, emotional intelligence, and accountability cannot be delegated to machines. AI can accelerate thinking, but meaning-making, trust-building, and moral courage are the true leadership differentiators in an AI-saturated world.

Story: 13

The Work AI Can't Do: A Day in the Life of an AI-Agile Leader

> **"The real danger is not that machines will think like humans, but that humans will think like machines."**
>
> — *Sydney J. Harris*

Rohan's Day: Becoming Irreplaceable in the Age of AI

6:47 AM – The Wake-Up Call Rohan's phone buzzed: *"Your role has 73% automation probability in 3 years."* Coffee forgotten, he stared at the ceiling. Years of effort. Senior Release Manager. And now—a statistic.

Then a ping from **Aria**, his AI coach: *"Want to talk about what machines can never take from you?"* Rohan smiled. "Alright, let's do this."

7:00 AM – The Alignment Pulse

Rohan sat at his kitchen table, steam rising from his cup. Outside, Chennai was waking up—auto-rickshaws honking, vendors calling, the city's relentless hum.

Rohan: "Aria, be honest with me. If AI can write code, manage projects, and optimize workflows... what am I actually *for* anymore?"

Aria: "I've analyzed your last six months of performance data, 360 feedback, and impact metrics. May I share what I see?"

Rohan: "Hit me."

Aria: "Your execution scores are exceptional—9.2/10. You deliver on time, within budget, with quality. But here's what's interesting: your highest-rated contributions aren't the deliverables. They're the moments when you..."

She paused deliberately.

Rohan: "When I what?"

Aria: "When you *reframed impossible situations*. When you *built trust across hostile teams*. When you *saw opportunities hidden in constraints*. What three platform crises did you resolve last quarter? My analysis shows you succeeded not through technical prowess, but through **contextual wisdom**—understanding what the data couldn't tell you about organizational politics, individual motivations, and timing."

Rohan leaned back. "So, you're saying..."

Aria: "Your automation risk isn't about your technical skills. It's about whether you're developing the skills that have *zero* automation probability: **Cognitive Agility, Emotional Intelligence, Creative Synthesis, Ethical Judgment, Systems Thinking, and Existential Leadership.**"

She displayed a simple graph on his screen:

SKILLS ASSESSMENT: Rohan Kumar

Technical Execution 92% [HIGH AUTOMATION RISK]

Data Analysis 85% [HIGH AUTOMATION RISK]

Cognitive Agility 58% [DEVELOPABLE ADVANTAGE]

Emotional Intelligence 74% [STRONG, CAN BE STRONGER]

Creative Synthesis 51% [UNTAPPED POTENTIAL]

Contextual Judgment 76% [YOUR HIDDEN SUPERPOWER]

Rohan: "So what's today's focus?"

Aria: "You have a perfect opportunity today. The Nexus client situation—the one keeping you up at night. That's not a technical problem. It's a **cognitive agility** challenge disguised as a compliance issue. Ready to practice problem reframing at the grandmaster level?"

Rohan: "Show me how."

8:15 AM – The Impossible Request Client Vikram demanded a real-time data export, violating policy. AI suggested denial. **Rohan:** "Perfectly logical, completely tone-deaf. Aria, the real problem isn't the export—it's trust and perception."

He reframed:

- Surface: client needs data export
- Hidden: client must show board control
- Deeper: fragmented internal systems
- Root: pressure to prove ROI

Solution options:

1. Temporary monitored exception
2. Read-only dashboard (compliant, elegant)
3. Pilot API bridge for innovation

He chose the dashboard—human judgment amplified by AI insight.

10:30 AM – Team Crisis AI workflow adoption failed. Team morale low.

Aria: "Identify emotions first." Rohan recognized: frustration, fear, threat.

Strategy:

1. Listening sessions
2. Reframe AI as a cognitive ally
3. Model vulnerability

The AI was perfect technically—but only human empathy could restore trust.

3:30 PM – The Creative Breakthrough

Rohan needed to redesign the AI workflow training. The current version was 47 slides of screenshots and bullet points. Deadly boring.

Rohan: "Aria, help me make this training not suck."

Aria: "Let's activate creative synthesis. Give me two completely unrelated concepts."

Rohan: "Uh... workflow training and... vintage video games?"

Aria: "Perfect. Now build a metaphor."

Rohan's mind started connecting dots:

"What if we gamified the entire onboarding? Call it ***'The AI Ally Quest.'*** *Employees are heroes. Each workflow mastery is a level-up. Compliance*

checks are 'Boss Battles.' The AI assistant is their sidekick—like Navi in Zelda—annoying sometimes but ultimately helpful."

He got excited. Really excited.

"We create achievement badges. We have a leaderboard for 'Most Creative AI Prompt.' We make it fun, competitive, and collaborative. We turn anxiety into adventure."

Aria: "This is genuine creativity. I could generate 1,000 training formats. But I couldn't create *meaning* and *emotional engagement* through metaphor. That's pure human magic."

5:45 PM – The Partnership Principle

As the day wound down, Rohan reflected on what had changed.

Rohan: "Aria, I need to ask you something uncomfortable."

Aria: "Go ahead."

Rohan: "Am I becoming dependent on you? Like, am I outsourcing my thinking?"

Aria: "That's the right question to ask. Here's my analysis: Over the past 90 days, you've made 247 significant decisions. In 89% of cases, you disagreed with my initial suggestion and chose a different path. Your decision quality score improved by 34%, and your team satisfaction improved by 28%."

Rohan: "So what does that mean?"

Aria: "It means you're using me correctly. I'm not replacing your judgment—I'm **stress-testing** it. I give you data, patterns, and possibilities. You provide context, wisdom, and values. The final decision is always yours."

Rohan: "So the rule is: you handle the cognitive heavy lifting—research, analysis, pattern recognition. I handle the synthesis, judgment, and relationship navigation."

Aria: "Precisely. **Intelligence Amplification, not Replacement.** You're not AI-dependent. You're AI-agile."

6:30 PM – The Evening Reflection

Rohan walked home through Chennai's evening chaos. The city never stopped. Neither could he.

But something had shifted.

This morning, he'd felt threatened by automation.

Tonight, he felt... capable. **Irreplaceable.**

Not because he could outwork machines.

But because he could:

- **Reframe problems** machines saw as fixed
- **Build trust** where algorithms created fear
- **Create meaning** where data offered only patterns
- **Navigate ambiguity** with wisdom and courage

He texted Meera: *"I think I figured out what I'm optimizing for. Not efficiency. Humanity."*

She replied: *"Now you're getting it.*

Automation risk dropped to 12%—not from technical skill, but from developing irreplaceable human capacities:

- Cognitive Agility
- Emotional Intelligence
- Creative Synthesis
- Ethical Judgment
- Systems Thinking
- Existential Leadership

Lesson: Don't compete with AI. Dance with it. Amplify your humanity.

Your Choice: Optimize for efficiency—and become replaceable. Or cultivate humanity—and become irreplaceable.

Here are the **top 3 key takeaways**:

- **AI replaces tasks, not human judgment**

  What makes leaders irreplaceable is not execution or analysis, but the ability to reframe problems, read context, build trust, and make ethical decisions where data alone is insufficient.

- **AI Agility means partnership, not dependence**

  The most effective leaders use AI for cognitive heavy lifting (data, patterns, options) while retaining ownership of synthesis, values, relationships, and final decisions.

- **Human capabilities are the true competitive advantage**

  Skills like cognitive agility, emotional intelligence, creative synthesis, and systems thinking have near-zero automation risk—and directly increase impact, trust, and leadership effectiveness.

Story: 14

The Human Side of AI Success: How Leaders Make AI Stick Every Day

> **"Leadership is practiced not so much in words as in attitude and in actions."**
>
> — *Harold S. Geneen*

AI transformation doesn't happen in strategy reviews.. It happens in **daily interactions**—in meetings, decisions, questions, and behaviors that leader's model.

Leadership doesn't fail because strategies are wrong.

It fails because strategy never makes it into daily behavior.

AI is no different.

What actually turns an AI strategy into lived reality is not more tools, roadmaps, or dashboards—it's what leaders do, say, and model every single day. Small, human actions quietly remove friction, build trust, and allow intelligent systems to take root.

Here's what that looks like in practice.

Begin with purpose, every time.

When leaders start AI conversations by asking *why* instead of *what*, the energy in the room shifts. The focus moves away from technology and toward real human or business problems. This simple habit protects teams from chasing novelty and keeps everyone anchored in value.

Model learning, not mastery.

The most powerful phrase a leader can use in an AI-driven world is, *"I don't know yet—let's figure it out together."* When leaders say this out loud, fear softens. Curiosity rises. Teams stop performing and start doing real work.

Make space for uncertainty without rushing to fix it.

When leaders ask what feels unclear, risky, or uncomfortable—and then truly listen—they create psychological safety. Concerns surface early. Hidden risks become visible. What could have turned into silent resistance becomes shared understanding.

Translate progress into meaning.

Numbers motivate for a moment. Meaning motivates for the long run. When leaders regularly connect AI work to real human impact—how it helped a customer, a colleague, or a process—teams feel their effort matters. Momentum grows naturally.

Slow down the moment before a decision.

AI accelerates insight. Leaders must slow interpretation. A brief pause—asking whether the team is reacting to pressure or responding to evidence—often prevents costly missteps. Better decisions come from clarity, not speed alone.

Treat ethics as a daily reflex.

Ethics shouldn't live in documents or quarterly reviews. When leaders routinely ask who might be negatively affected by a decision, ethical judgment becomes part of normal thinking. Over time, responsibility turns into a habit.

Invite voices that don't usually speak.

Great AI outcomes rarely come from narrow perspectives. By intentionally bringing in people from different roles or disciplines, leaders surface blind spots early. Assumptions are challenged before they harden into mistakes.

Protect energy as fiercely as performance.

Learning and innovation collapse under exhaustion. Leaders who regularly ask what can be simplified or dropped send a powerful signal: sustainability matters. Teams stay sharp because they're not constantly overwhelmed.

Turn learning into action.

Reflection only matters if it changes behavior. When leaders ask what the team learned—and how that learning will shape next week's actions—progress compounds. Insight becomes evolution.

Show up, especially when things are unclear.

Leadership presence matters most when answers aren't obvious. Consistent visibility during uncertainty builds trust faster than flawless execution ever could. People don't need perfection—they need steadiness.

The deeper truth

You don't need to transform your organization overnight.

You need to transform how you show up—daily.

When leaders change their behaviors, systems follow.

And that's how AI stops being a strategy—and becomes the way work actually happens.

AI success is built one conversation, one question, and one behavior at a time.

And when leaders model clarity, curiosity, courage, and care, teams follow.

Here are **3 clear, practical key takeaways** leaders can implement daily:

- **AI sticks when leaders model behavior, not just strategy**

  AI adoption succeeds in everyday moments—how leaders ask questions, admit uncertainty, pause before deciding, and show up consistently. What leaders do daily matters more than what they announce quarterly.

- **Psychological safety and clarity unlock real AI adoption**

  When leaders start with *why*, create safe spaces for uncertainty, and connect AI work to human impact, fear drops, and learning accelerates. Trust and meaning turn AI from a tool into a team capability.

- **Small reflective habits outperform big transformation plans**

  Simple practices—pausing before decisions, inviting diverse voices, closing learning loops, and making ethics routine—compound over time and make AI adoption sustainable rather than performative.

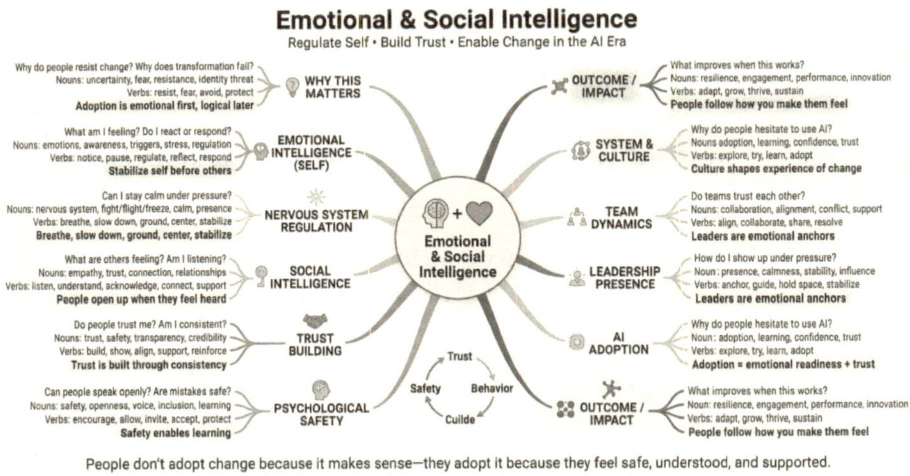

Top 5 Key Takeaways — Emotional & Social Intelligence in the AI Age

- **Emotional intelligence is the foundation of AI leadership**

  Before leaders can guide others through AI-driven change, they must regulate their own fear, uncertainty, and identity threats. Self-awareness comes before strategy, tools, or intelligence.

- **AI challenge's identity more than capability—and leaders must name that reality**

  Resistance to AI is rarely technical. It's emotional. When leaders separate self-worth from expertise, they unlock learning, adaptability, and trust—both for themselves and their teams.

- **AI must amplify human judgment, not replace it**

  The most effective leaders treat AI as a thinking partner, not an answer machine. AI handles data and patterns; humans own context, ethics, meaning, and accountability.

- **What remains human determines long-term advantage**

  Skills with near-zero automation risk—emotional intelligence, cognitive agility, creative synthesis, systems thinking, and moral courage—are what make leaders irreplaceable in an AI-saturated world.

- **AI adoption succeeds through daily behavior, not grand plans**

  AI becomes real through everyday leadership habits: modelling learning, creating psychological safety, pausing before decisions, making ethics routine, and connecting AI work to human impact. Small, consistent behaviors compound into lasting transformation.

Part IV

N — Neuro-Adaptive Learning

Learning Faster Than Change

Introduction

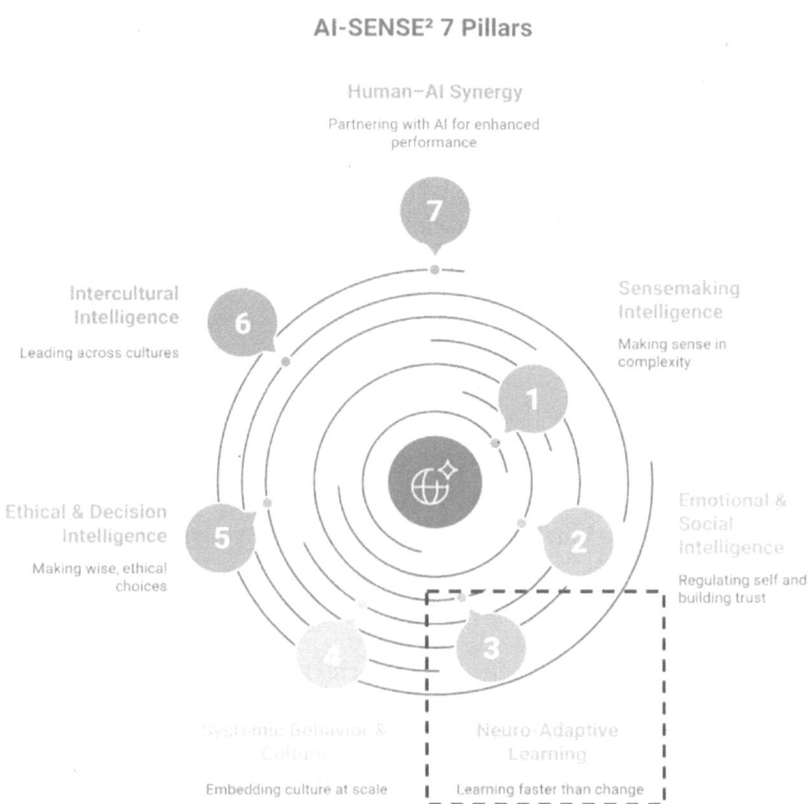

The leaders who will thrive in the age of artificial intelligence will not be the fastest, the most technical, or the most confident.

They will be the ones who can **learn faster than change**, **think more deeply than the algorithms**, and **remain emotionally steady while the world accelerates around them**.

This chapter marks a turning point.

Up to now, leadership success was built on expertise—on knowing more, deciding faster, and projecting certainty. But in a world where AI updates itself continuously, produces instant insights, and challenges human judgment at scale, expertise decays quickly. What replaces it is something far more demanding—and far more human.

Learning becomes the new leadership currency.

Not episodic learning. Not training programs. But continuous, neuro-adaptive learning—the ability to update assumptions, regulate fear, unlearn success patterns, and evolve thinking in real time. Leaders who cling to mastery fall behind. Leaders who embrace curiosity stay relevant.

At the same time, AI introduces a paradox. While machines accelerate information, they also amplify emotional pressure. Alerts trigger urgency. Predictions provoke anxiety. Confidence scores invite blind trust. Without emotional clarity, leaders mistake speed for insight—and reaction for leadership.

That is why **emotional intelligence becomes the control system of AI-era leadership**.

This chapter explores why leadership presence matters more—not less—when AI enters decision-making. Why slowing down interpretation is the only way to regain clarity. Why Flow and deep thinking are no longer luxuries, but strategic necessities. And why the most powerful AI systems fail unless the human systems around them are ready.

Across these pages, you will see a clear pattern emerge:

- AI accelerates data—but humans must create meaning
- AI optimizes efficiency—but humans must protect ethics and trust

- AI learns endlessly—but humans must learn wisely
- AI scales intelligence—but humans scale impact

Whether you are leading people, products, or complex innovation programs, this chapter invites you to step into a different posture of leadership—one grounded in curiosity, emotional regulation, deep thinking, and deliberate learning.

Because in the end, the future will not belong to leaders who know the most.

It will belong to those who can **stay present, keep learning, and remain human—while everything else changes**.

Story: 15

Why Learning Is the New Leadership Currency: The Rise of Neuro-Adaptive Leaders

> **"The illiterate of the 21st century will not be those who cannot read and write, but those who cannot learn, unlearn, and relearn."**
>
> — *Alvin Toffler*

The Story of the Leader Who Realised Learning Had Changed

In a world where expertise now expires faster than technology itself, a leadership coach began telling a story that stayed with people long after his sessions ended.

He always started the same way.

At a private leadership offsite, he once asked a room full of senior executives a simple question.

"What skill got you promoted to your current role?"

The answers came quickly.

Strategic thinking.

Technical expertise.

Deep industry knowledge.

Years of experience.

The room was confident. These leaders had spent decades building their expertise.

Then the coach asked a second question.

"What skill will keep you relevant in this role for the next three years?"

This time, the room went quiet.

Not because the leaders lacked intelligence—but because they could feel the truth before they could explain it.

The expertise that brought them here would not be enough to take them forward.

The Knowledge Depreciation Crisis

The coach then shared a reality that unsettled even the most experienced executives.

The half-life of professional skills, he explained, is now estimated at around **five years**.

In technology-driven industries, it can be closer to **two and a half years**.

That means half of what leaders know today may become obsolete, automated, or irrelevant before their next major career milestone.

For decades, leadership followed a predictable formula:

Accumulate knowledge → Gain expertise → Earn authority → Lead with confidence

But that formula no longer worked.

Not because knowledge had lost its value.

But because knowledge now **expires faster than leaders can accumulate it.**

The real leadership advantage today is no longer mastery.

It is learning velocity.

The leaders who remain relevant are not those who know the most.

They are the ones who **learn faster than the world is changing.**

This was the beginning of what the coach called **Neuro-Adaptive Leadership.**

From Mastery to Adaptability

The most effective leaders, he explained, had already made a quiet but powerful shift.

The old leadership posture looked like this:

- Mastery: *"I've arrived. I know this."*
- Expertise: *"I'm the expert—listen to me."*
- Control: *"I manage outcomes through certainty."*

But the new leadership posture looked very different:

- Adaptability: *"I am always learning."*
- Curiosity: *"I ask better questions than I give answers."*
- Experimentation: *"I manage outcomes through learning quickly."*

This change was not just philosophical.

It was neurological.

Leadership effectiveness now depended on **how quickly the brain could update how it thinks.**

And that is difficult—because the human brain naturally resists change.

Why Change Feels Like a Threat

To make this real, the coach shared the story of a senior executive he called **Sarah**.

Sarah had built her career on operational excellence. She had an extraordinary ability to optimize complex systems. Over time, she developed an instinct for improving efficiency.

That expertise made her indispensable.

Then her organization introduced AI into operations.

Suddenly, algorithms were optimizing processes faster—and more accurately—than she ever could.

At first, Sarah resisted.

She worked harder.

She doubled down on manual processes.

She tried to prove the old methods still worked.

But for the first time in her career, effort stopped producing results.

From the outside, some people called her stubborn.

Others thought she was in denial.

But the coach explained what was really happening.

When familiar expertise is threatened, the brain interprets it as danger.

- Loss of competence threatens identity
- Loss of status threatens belonging
- Fear of irrelevance feels existential

The brain reacts with its oldest defense mechanism.

Fight. Flight. Or freeze.

Resistance to change, he explained, is rarely logical.

It is biological.

The Breakthrough

Sarah's turning point came when someone offered her a simple reframe.

She realized something powerful.

"I'm not losing my value," she said.

"I'm upgrading it."

Her operational expertise did not disappear.

It evolved.

Instead of competing with AI, she began learning how AI could enhance operations. Her experience gave her context that the algorithms did not have.

Slowly, fear gave way to curiosity.

And curiosity gave way to learning.

The Most Underrated Leadership Skill: Unlearning

At this point in the story, the coach would pause and ask the room another uncomfortable question.

What made you successful in the past that might now be limiting your future?

This, he explained, is **unlearning**.

And it is far harder than learning something new.

Success strengthens neural pathways. The brain becomes efficient at repeating what once worked.

But what worked yesterday can quietly limit relevance today.

He gave familiar examples:

- A sales leader relying only on relationships while ignoring data insights
- An engineer dismissing AI because "real professionals don't use shortcuts."
- An executive demanding an annual plan in markets that change quarterly

These leaders were not incompetent.

They were simply **over-invested in past success.**

Unlearning does not discard wisdom.

It updates it.

Three Questions Adaptive Leaders Ask

To stay relevant, the coach encouraged leaders to ask themselves three questions every quarter.

1. What assumptions am I still operating from?

"My team needs me to have all the answers."

Maybe they need better questions instead.

2. Which habits no longer serve the system?

"I personally review every decision."

At scale, that becomes friction, not leadership.

3. Where am I confusing experience with relevance?

"I've worked in this industry for twenty years."

That describes history—not necessarily the present.

Relearning: Changing How Leaders Think

Later, Sarah asked the coach a simple question.

"So, what should I learn now?"

The answer surprised her.

"The real question," he said, **"is not what you learn. It's how you learn."**

The world already has more information than any person can absorb.

What matters is **how leaders process information and adapt their thinking.**

Neuro-adaptive leaders learn in cycles.

They test small ideas.

Observe what happens.

Adjust quickly.

They learn through action, reflection, and feedback.

Mistakes are not treated as failure.

They are treated as **data**.

Because when people feel safe, the brain stays curious.

When people feel judged or afraid, learning shuts down.

Relevance Is a Practice, not a Destination

Relevance is no longer something leaders achieve once and keep forever.

In a fast-changing world, it must be earned again and again.

Leaders stay relevant by **learning as fast as the world changes.**

They stay curious.

They question their assumptions.

They continuously improve how they think—not just what they know.

AI can generate insights.

But human judgment still provides context, ethics, and meaning.

The strongest leaders combine both.

Technology and wisdom.

From Expert to Learning Architect

One of the hardest transitions for many leaders is giving up the need to be the smartest person in the room.

For years, their value came from having answers.

But in a rapidly changing world, no one person can know everything.

Adaptive leaders take on a different role.

They become **architects of learning environments.**

Instead of giving all the answers, they create spaces where teams can explore ideas, experiment, and learn together.

They reward curiosity.

They encourage questioning.

They model openness instead of certainty.

And by doing this, they create teams that learn faster than the problems they face.

The Story's Final Message

The coach always ended the story the same way.

Neuro-adaptive leadership is not about running faster forever.

It is about evolving without losing humanity.

It allows:

- Curiosity without anxiety
- Growth without exhaustion
- Adaptation without abandoning values

And the future, he reminded them, will not belong to the most confident leaders.

It will belong to the most **adaptive, reflective, and human** ones.

Because the most powerful words a leader can say today are no longer:

"I know."

But instead:

"I'm still learning. Join me."

Here are **the top 3 key takeaways**:

- **Learning velocity matters more than expertise**

  In a world where skills expire quickly, leadership advantage no longer comes from what you know, but from how fast you can learn, unlearn, and relearn. Relevance is sustained through adaptability, not mastery.

- **Resistance to change is biological, not intellectual**

  Leaders often resist AI and new ways of working not because they don't understand them, but because change threatens identity, status, and self-worth. Neuro-adaptive leaders acknowledge this fear, regulate it, and turn anxiety into curiosity.

- **The strongest leaders shift from expert to learning architect**

  Future-ready leaders stop trying to be the smartest person in the room. Instead, they model curiosity, ask better questions, create psychological safety, and design environments where continuous learning becomes the norm.

Story: 16

Why Learning Speed Is the New Leadership Advantage

> **"He who conquers himself is the mightiest warrior."**
>
> — *Confucius*

When Machines Move Fast, Leaders Must Learn Faster

At 7:45 in the morning, the office was still quiet.

Arjun, a senior leader responsible for large transformation programs, walked in with his coffee and opened his laptop.

Within seconds, the calm disappeared.

His screen filled with alerts.

AI dashboards highlighted potential risks.

Market signals flashed warnings.

Customer sentiment analysis suggested that something might be shifting.

His calendar was already packed with meetings.

It felt as if the digital world had started running faster than he could keep up.

Inside his body, he felt it immediately.

His chest tightened.

His heartbeat faster.

A small voice inside whispered:

"Act now. Decide quickly. Don't fall behind."

For a moment, the pressure felt overwhelming.

But Arjun paused.

He had seen this pattern before.

The Speed–Anxiety Loop

In the AI era, leaders often fall into what Arjun called the **speed–anxiety loop**.

First, AI produces insights at extraordinary speed—signals, patterns, and predictions faster than the human mind can fully absorb.

Then emotions rise.

Anxiety appears.

Fear of missing out grows.

The pressure to react immediately becomes intense.

When leaders respond to this pressure, decisions often come from stress rather than clarity.

And in those moments, the emotional noise becomes more dangerous than the data itself.

Arjun took a deep breath.

He reminded himself of something he had learned over time.

“AI makes information faster.

Only humans can slow down meaning.”

Creating Space to Think

Instead of reacting immediately, Arjun opened a notebook and drew two columns.

On one side, he wrote:

FACTS

On the other side:

FEELINGS

He calmly listed the signals the AI system had detected.

Then he wrote down his own emotional reactions to those signals.

Within minutes, the tension in his body began to fade.

By separating facts from feelings, he regained something that AI could never provide:

the ability to pause, reflect, and think clearly.

But something else happened in that quiet moment.

He began to see the signals differently.

The alerts were not simply warnings.

They were **learning signals**.

Turning Signals into Learning

Instead of asking, “How do we fix this quickly?”

Arjun asked a different question:

“What is this system trying to teach us?”

The data suggested that customer behavior was beginning to shift.

Not dramatically.

But enough to signal that expectations were changing.

What looked like a problem might actually be an early opportunity.

In that moment, Arjun realized something important.

Leadership in the AI era was no longer about reacting faster.

It was about **learning faster.**

The leaders who stay relevant are not those who know the most.

They are the ones who **turn signals into learning before others do.**

The Courage to Unlearn

As Arjun reviewed the insights more carefully, another realization appeared.

The strategy his team had been using for months was beginning to show weaker results.

It had worked well in the past.

But the market had started evolving.

For a moment, Arjun felt resistance.

Success had reinforced those methods. They felt reliable.

But he knew something important.

What once created success can quietly become the barrier to the next level of growth.

Leadership now required something difficult.

Not just learning.

But **unlearning.**

Letting go of assumptions that no longer matched reality.

Updating how the team thought about the problem.

Instead of defending past strategies, Arjun decided to explore new possibilities.

A Team That Learns Together

Later that day, Arjun gathered his team to discuss the signals.

He entered the room calm and focused.

Instead of saying,

“We need to fix this immediately.”

He said something different.

“Let’s understand what this signal is teaching us.”

The tone of the room changed instantly.

People stopped reacting and started thinking.

Instead of defending their ideas, they began exploring possibilities.

Different perspectives emerged.

Marketing shared insights about shifting customer expectations.

Product teams discussed emerging usage patterns.

Data scientists explored additional trends the AI system had detected.

What initially looked like a problem became something more valuable.

A learning moment for the entire team.

Together, they realized the signal was not a crisis.

It was an early indicator of a new customer need.

And because they noticed it early, they had time to respond before competitors did.

Learning That Spreads

The insight did not stop with Arjun’s team.

He shared the discovery across other departments in the organization.

Soon, other teams began examining similar signals.

Small adjustments were made across multiple products and services.

What began as a single AI alert evolved into a broader organizational insight.

The company had not just reacted to data.

It had **learned from it.**

And that learning strengthened the entire system.

The Real Future of Leadership

Later that evening, Arjun reflected on the day.

AI would continue to grow faster.

Data would continue expanding.

Signals would appear more frequently and more quickly.

But leadership would still depend on something deeply human.

The ability to pause.

The ability to reflect.

The ability to learn.

Machines could generate information at incredible speed.

But **learning still belonged to humans.**

And the leaders who stayed relevant would not be those who reacted the fastest.

They would be those who **learned the fastest.**

The Leadership Identity Shift

For many years, leadership meant having the answers.

Experience created authority.

Knowledge created confidence.

But the AI era had changed that identity.

The most effective leaders were no longer those who said:

"I know."

They were the ones who said:

"Let's learn."

Because in a world where machines learn every second, the leaders who succeed will be those who learn just as quickly.

And who help their teams and organizations learn with them.

The Real Advantage

In the AI era, competitive advantage no longer comes from what leaders know.

It comes from **how quickly they can learn, unlearn, and adapt.**

Because the future will not belong to the most confident leaders.

It will belong to the most curious ones.

The ones who turn signals into insight.

Insight into learning.

And learning into progress.

Here are the **top 3 key takeaways**:

- **Learning Faster Than Change Is the New Leadership Advantage**

  In the AI era, knowledge expires quickly. What leaders knew yesterday may not work tomorrow.

  Because of this, leadership advantage no longer comes from accumulated expertise. It comes from the ability to **learn, unlearn, and adapt faster than the environment changes.**

  Great leaders do not try to know everything.

  They build the ability to **continuously update how they think.**

- **AI Generates Signals, but Humans Create Meaning**

  Artificial intelligence can detect patterns, risks, and opportunities faster than humans. However, AI does not understand context, values, or long-term consequences.

  Leaders must interpret what the signals mean.

  AI shows **patterns and probabilities**, but humans provide **judgment, wisdom, and direction.**

  The real power comes from the partnership between **machine intelligence and human understanding.**

- **Great Leaders Build Learning Systems, Not Just Solutions**

  The most effective leaders do not simply react to information. They turn signals into learning.

  They help:

  - individuals update their thinking
  - teams experiment and learn together
  - organizations spread insights across the system

  In this way, leadership shifts from **being the person with answers** to **being the architect of continuous learning.**

Story: 17

Flow in the Age of AI: How Leaders Think Deeply While Algorithms Move Fast

> **"The real problem of humanity is the following: we have Paleolithic emotions, medieval institutions, and god-like technology."**
>
> — *E. O. Wilson*

It was 9:00 a.m., and Jaya, a product leader at a fast-growing tech company, stared at her screen. Her inbox was overflowing, notifications pinged every few seconds, and AI-generated insights suggested ten different strategies for her upcoming board meeting.

She felt the familiar tug of overwhelm—the modern leader's invisible enemy.

And yet, she remembered what she had learned about **AI Agility**: in a world dominated by algorithms, speed, and endless data, the only thing that could cut through the noise was **human deep thinking—the Flow State**.

The Hidden Power of Flow

Flow is more than focus.

It's a state where your brain locks onto a challenge, and time itself seems to bend.

Mihaly Csikszentmihalyi described it as complete absorption, but neuroscience explains it in even more exciting terms.

Dopamine sharpens attention and curiosity, noradrenaline filters distractions, and repeated engagement strengthens myelin in the neural pathways—making your thinking faster, smarter, and more fluid.

For Jaya, this meant something radical: she could treat AI not as a crutch, but as a partner. The AI tools around her could analyze patterns, generate drafts, and summarize trends, but only she could synthesize them into a **strategic, ethical, and creative vision**.

AI Thinking Meets Human Flow

This is where **AI Agility** comes alive. AI excels at speed, data crunching, and pattern recognition—essentially acting as a supercharged System 1. But the breakthroughs, the truly transformative decisions, require **System 2 thinking**: deliberate, reflective, deeply human.

- AI flags anomalies in sales trends, but Jaya connects the dots with customer empathy.
- AI generates hundreds of product iterations, but she selects the path aligned with long-term vision.
- AI predicts market shifts, but she evaluates the ethical, social, and cultural impact.

Without her Flow State, she would be reacting to notifications, drowning in shallow thinking. With it, she directed the AI, not the other way around.

Applying Flow Across Teams

AI Agility isn't only for leaders—it transforms teams too. Imagine a team of analysts using AI for data prep while they enter deep collaborative Flow for strategic synthesis:

- **Shared Flow Blocks**: Teams set dedicated hours for deep collaborative work, free from messaging apps and distractions.
- **AI-Powered Support**: AI handles repetitive analysis, letting the human team focus on insights, ideation, and ethical decision-making.
- **Micro-Feedback Loops**: Quick, real-time reflections keep the team aligned, sharpening both individual and collective cognition.

In one week, Jaya's team not only delivered a more robust strategy but did so with clarity, creativity, and less stress—a perfect example of human-AI symbiosis in action.

Four Keys to Cultivating AI-Enhanced Flow

1. **Prioritize High-Leverage Work**: Identify the tasks that require uniquely human judgment and creativity. Let AI handle the repetitive, data-heavy work.

2. **Create Flow Sanctuaries**: Protect 90–120 minutes of uninterrupted time for deep thinking. Turn off alerts, silence devices, and remove distractions.

3. **Set Clear Goals with Immediate Feedback**: Break complex challenges into actionable micro-goals. Dopamine thrives on small wins.

4. **Resist Low-Value Distractions**: Delay shallow work until after deep work sessions. Train your brain to crave depth over speed.

In the age of AI, true leadership isn't about keeping up with the fastest algorithm. It's about **mastering your own brain**, directing AI thoughtfully, and entering the Flow State where insight, innovation, and ethical judgment intersect.

Because in a world of data, notifications, and predictions, the leaders who will thrive are those who combine **AI thinking with human deep thinking**—the essence of **AI Agility**.

Here are the **Top 5 Daily Actions**

To truly thrive in an era where technology moves faster than we can think, we have to change the way we approach our workdays. AI Agility isn't just a concept; it's a daily practice of choosing where to spend your most valuable resource: your attention.

The first step is a radical shift toward **high-leverage work**. This means looking at your to-do list and identifying the tasks that only you can do—those that require human judgment, creative spark, or ethical nuances. Everything else, like crunching massive data sets or managing repetitive schedules, should be handed over to the machine. By letting AI handle the "drudge work," you clear a path for the deep, meaningful contributions that define true leadership.

To protect this path, you must build **Flow Sanctuaries**. These are protected blocks of time, ideally 90 to 120 minutes each day, where you engage in deep work. During this time, the digital world is silenced; notifications are turned off, and tabs are closed. This isn't just about focus; it's about giving your brain the quiet it needs to reach its peak performance. To stay motivated during these sessions, it helps to break massive projects into **micro-goals**. Each small victory provides a quick hit of dopamine, turning the grind into a series of wins that keep you moving forward.

A critical part of this discipline is the ability to **resist shallow distractions**. In a world of instant pings, it is easy to spend the whole day reacting to emails and messages. However, high-impact leaders delay this "shallow work" until their deep work is finished. They train their minds to crave the satisfaction of sustained attention rather than the quick, hollow thrill of multitasking.

Ultimately, this leads to a powerful form of **Human-AI Collaboration**. Instead of seeing AI as a replacement, you treat it as a brilliant partner. You use it to find patterns and generate initial drafts, but you don't stop there. You take those insights and apply your own empathy, experience, and ethical reasoning to turn them into a strategy. This blend of machine speed and human depth is where true clarity and creativity live.

Here are the **top 3 key takeaways**:

- **AI creates speed—but Flow creates insight.**

  AI excels at rapid analysis and pattern detection, but real leadership value comes from human deep thinking. Leaders must intentionally enter Flow states to synthesize data into strategy, ethics, and meaning instead of reacting to endless alerts.

- **The leader's edge is directing AI, not keeping up with it.**

  When leaders use AI for cognitive heavy lifting and reserve their own energy for judgment, empathy, and long-term vision, AI becomes a partner—not a distraction or a crutch.

- **Flow is a discipline, not a luxury, in the AI age.**

  Protecting uninterrupted deep work, prioritizing high-leverage human tasks, and resisting low-value distractions are daily leadership practices that turn AI speed into clarity, creativity, and sustainable performance.

Story: 18

The AI Learning Mindset: How Product Teams Can Master AI Agility

> **"In times of change, learners inherit the future, while the learned find themselves beautifully equipped to deal with a world that no longer exists."**
>
> — *Eric Hoffer*

AI is no longer a separate initiative managed by a niche data science unit.

It now flows through every layer of product work—from discovery conversations and design reviews to coding, testing, feature rollout, and user feedback loops.

In this fast-evolving landscape, the rhythm of learning must change. Annual training sessions or static onboarding documents cannot keep pace with the real-time evolution of AI tools and methodologies.

To thrive in the AI-first era, software product teams need something deeper and far more dynamic: **the AI Learning Mindset**.

Built on *Curiosity, Micro-learning,* and *Neuroplasticity*, this mindset becomes the fuel for true AI Agility—the ability to learn, adapt, and innovate faster than the speed of technological change.

Curiosity: The Engine of Innovation and Better Decisions

For a modern product team, curiosity is not a soft skill; it is a strategic advantage. In an environment where AI can suggest system architectures, write test cases, simulate user flows, and accelerate coding, curiosity shifts the team from being passive recipients of AI outputs to active co-creators with AI systems.

Curiosity begins when the team consistently wonders: *How could AI help us do this faster? What if we tried a different model? What would happen if an AI assistant redesigned this workflow?* When these questions enter sprint planning, daily scrums, backlog refinement, and retrospectives, exploration becomes part of the team's identity.

Product managers start looking for patterns in user behavior that AI can surface. Designers begin inviting generative models to reimagine journey maps and visual explorations. Developers experiment with copilots during prototyping. Even test engineers harness AI to detect anomalies or predict defects. Over time, curiosity transforms into a culture—one where new tools are explored without hesitation, experiments are welcomed, and fresh ideas become part of everyday work rather than occasional innovation spikes.

This mindset helps teams discover efficiencies, uncover non-obvious insights, and create solutions that would never emerge within the boundaries of traditional thinking.

Micro-Learning: Building AI Capability into the Daily Workflow

In most software teams, schedules are packed, deadlines feel tight, and disruptions come quickly. Long training workshops often feel unrealistic—and by the time they arrive, the tools have already evolved.

Micro-learning becomes the answer: short, targeted learning experiences that integrate into the team's daily rhythm.

Instead of reserving hours or days for training, micro-learning turns each day into a small but consistent learning opportunity. A few minutes in a standup can become a spark for understanding a new AI prompt pattern.

A short video shared on Slack can reveal a breakthrough capability in a testing platform. Pair programming with an AI assistant turns into a daily lesson in prompting, reviewing, validating, and improving output.

Teams begin to embed these tiny learning moments into sprint goals—just one new AI technique per cycle, one improved prompt, or one automated task. Small, continuous steps start compounding into meaningful mastery. And because micro-learning blends directly into the existing workflow, it never becomes a burden. It feels like a natural extension of the workday—lightweight, frictionless, and immediately useful.

This approach builds AI fluency gradually but powerfully, allowing teams to stay current without ever feeling overwhelmed.

Neuroplasticity: Rewiring the Brain for AI-First Thinking

AI dominance doesn't just demand new skills—it requires new mental models. Teams must think in terms of data, not assumptions. They must learn to collaborate with AI as a cognitive partner rather than a distant tool.

They must grow comfortable with automation, probabilistic answers, and non-linear thinking. This shift calls on the brain's incredible ability to rewire itself: **neuroplasticity**.

When teams invite AI to challenge their assumptions, they begin stretching old thought patterns. An AI-generated alternative design compels a UX designer to consider new directions.

A model suggesting a simplified system architecture forces developers to rethink complexity. An AI analysis of test failure patterns teaches QA to see trends they might have missed.

Deliberate practice strengthens these new neural pathways. Repeatedly experimenting with prompting, fine-tuning workflows, interpreting AI results, and cross-collaborating across roles creates cognitive flexibility. Over time, the team becomes more adaptable, more inventive, and more willing to embrace rapid changes. Reflecting on these shifts in retrospectives reinforces the new mental habits.

Neuroplasticity makes the team not just skilled, but mentally prepared for the AI-driven future—a future defined by constant evolution.

The Emergence of the AI-Ready Product Team

When curiosity fuels exploration, micro-learning ensures continuous improvement, and neuroplasticity strengthens adaptability, the transformation is profound.

The team becomes faster because AI accelerates coding, testing, prioritization, and analysis. They become smarter because each day adds a small piece of knowledge that compounds into expertise. They grow more innovative because curiosity unlocks bold new ideas and unconventional solutions. And they become truly adaptable because their minds are conditioned to evolve with every new tool, every new model, and every emerging challenge.

The cultural shift is unmistakable. The early anxiety— *"AI might replace us"*—fades away. In its place emerges a new belief: *"AI helps us build better, faster, smarter products than ever before."*

This is AI Agility in action. And this is the mindset that will define the next generation of high-performance product teams.

Here are the **top 3 key takeaways**:

- **AI Agility is a learning capability, not a technical upgrade.**

  Product teams succeed with AI when they shift from static training to a continuous learning mindset. Curiosity, micro-learning, and adaptability matter more than mastering any single tool or model.

- **Curiosity turns AI from a tool into a co-creator.**

  Teams that regularly ask *why, how, and what if*—in sprints, standups, and retrospectives—use AI to uncover patterns, test ideas, and design better solutions rather than passively accepting AI outputs.

- **Small daily learning compounds into long-term advantage.**

  Micro-learning embedded into everyday work rewires how teams think. Over time, neuroplasticity builds cognitive flexibility, helping teams adapt faster, innovate continuously, and stay relevant as AI evolves.

Summary: Daily Habits to Build a Learning Mindset

In today's AI-driven world, things are changing very fast. Technology, tools, and skills are constantly evolving. Because of this, the most important ability is not just **what you know**, but **how quickly you can learn and adapt**.

Our brain has a special ability called **neuroplasticity**. This means the brain can change, grow, and form new connections when we learn new things. Just like muscles become stronger through exercise, the brain becomes stronger when we practice learning regularly.

This chapter explains that becoming a strong and adaptable leader does not happen in a training program or workshop. It happens through **small daily habits** that help you think better, learn faster, and stay curious.

Daily Habits for Individuals

Good leaders start their day by questioning their assumptions. Every morning, they ask themselves a simple question:

"Is there something I believe that might not be true anymore?"

This helps them keep an open mind.

At the end of the day, they reflect on their experiences. They write down:

- One thing they learned
- One thing that surprised them
- One idea or belief that was challenged

This small habit helps turn everyday experiences into real learning.

Leaders also try to step outside their comfort zone. A few times a week, they spend some time learning something new, such as an AI concept, a new tool, or a different way of solving problems.

Another important habit is protecting time for **deep focus**. When people work without distractions for a period of time, their thinking becomes clearer and more creative.

Finally, leaders stay curious about new technology. Trying a new AI tool or exploring a new idea each week helps them see change as an opportunity instead of a threat.

Learning Habits for Teams

Learning should not happen only at the individual level. Teams also need to learn together.

Teams can do this by asking simple questions regularly:

- What new thing did we try this week?
- What surprised us?
- What will we do differently next time?

This helps teams improve step by step.

Leaders should also create a safe environment where people are not afraid to try new ideas. When experiments do not work, they should be treated as **learning experiences**, not failures.

Teams also benefit from sharing ideas from outside their field, such as insights from books, articles, other industries, or new technologies.

Learning Habits for Organizations

Organizations grow stronger when knowledge is shared. After finishing a project, teams should record what they learned so others can benefit from their experience.

Leaders should also share what they are learning with others. When leaders openly talk about learning new things, it encourages everyone in the organization to keep learning as well.

The Most Important Daily Habits

Even three small habits can make a big difference:

- In the morning, question one assumption.
- During the day, focus deeply on an important task.
- In the evening, write down something you learned.

A New Way of Thinking

To succeed in the AI world, people must change how they see themselves.

Instead of thinking, **"I am someone who knows many things,"**

They should think, **"I am someone who is always learning."**

Knowledge can become outdated quickly, but the ability to learn will always be valuable.

People who build this habit of continuous learning will not only survive in the AI age—they will **lead and shape the future.**

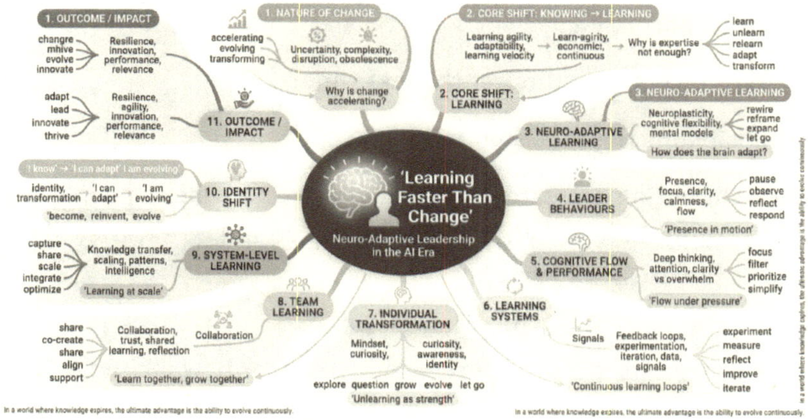

Top 5 Key Takeaways from These Chapters

- **Learning speed—not expertise—is the new leadership currency**

  Across leaders, teams, and organizations, relevance no longer comes from what you already know. It comes from how quickly you can **learn, unlearn, and relearn** as conditions change. Neuro-adaptive leaders stay relevant by updating their thinking faster than the environment evolves.

  Shift: From "I know" → to "I'm still learning."

- **AI amplifies speed, but humans must provide clarity and meaning**

  AI excels at fast signals, patterns, and probabilities. But without **System 2 thinking**, emotional regulation, and contextual judgment, speed turns into noise. The leaders who win are those who **slow down interpretation** even as AI speeds up information.

  Rule: AI accelerates data. Humans decide direction.

- **Emotional intelligence is not optional—it is the control system**

  Fear, identity threat, and anxiety appear *before* logic during AI change. Leaders who cannot regulate their inner state unintentionally spread panic, resistance, or blind automation bias. Emotional clarity becomes the **foundation of trust, presence, and sound judgment**.

  Truth: Unregulated emotion is the hidden failure mode of AI adoption.

- **Flow and deep thinking are strategic advantages in the AI age**

  In an always-on, alert-driven world, **Flow is no longer a luxury—it is a leadership discipline**. Leaders and teams who protect deep work, prioritize high-leverage human tasks, and direct AI intentionally produce better strategy, creativity, and ethical outcomes.

Edge: Don't keep up with AI—*direct it from depth.*

- **AI success depends on human systems being ready first**

  Whether in product teams or regulated manufacturing, AI breakthroughs fail without **Innovation Agility**—collaboration, resilience, curiosity, energy, and persistence. Technology scales only when **human trust, learning culture, and creativity** are designed deliberately.

  Bottom line:

  AI works when people are ready to work *with* it.

  One-Sentence Chapter Essence

  The future belongs to leaders and teams who learn faster than change, regulate emotion before reaction, think deeply while machines move fast, and design human systems that let AI amplify—not replace—judgment, creativity, and responsibility.

Knowledge Tree Guidebooks

Part V

S — Systemic Behavior & Culture

Making AI Stick at Scale

Introduction

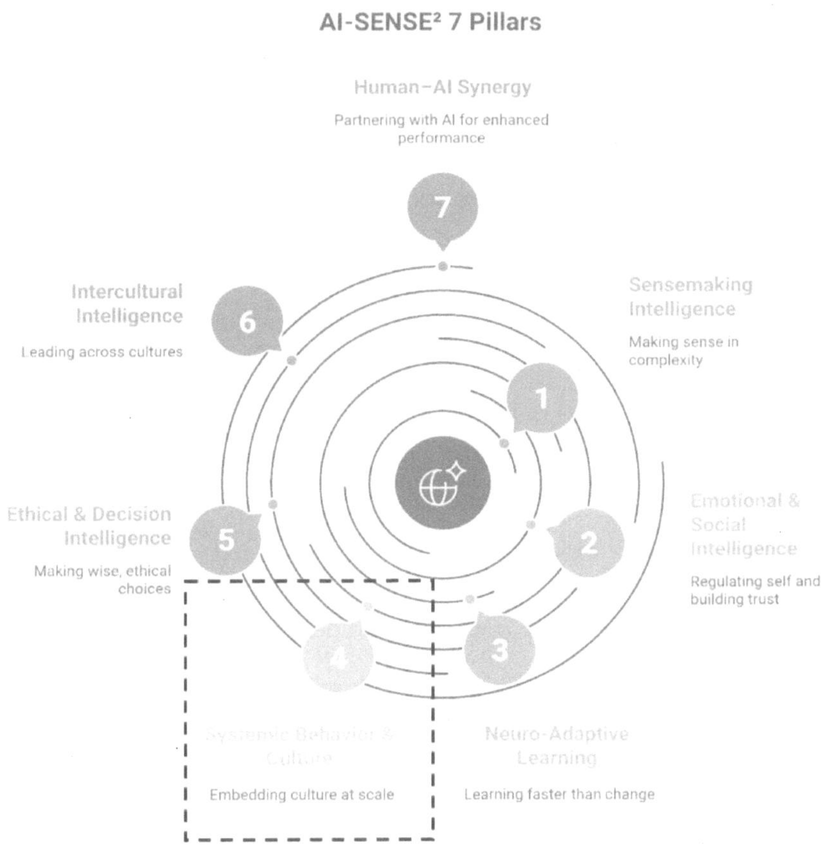

Many organizations succeed with AI pilots.

The models perform.

The technology works.

Budgets are approved.

And yet, when organizations attempt to scale AI across the enterprise, progress stalls quietly.

Part V—**Systemic Behavior & Culture** explains why.

This section shifts the focus from individual leadership capability to the broader **organizational system in which AI must operate**. It reveals a difficult reality of the AI era: no matter how advanced the technology becomes, intelligent systems cannot create lasting value unless the human system surrounding them evolves as well.

Culture, incentives, decision norms, and leadership expectations either reinforce intelligent behavior or silently prevent it.

Across these chapters, readers discover why AI rarely fixes underlying organizational problems. Instead, it **magnifies what already exists**. When judgment is weak, AI accelerates poor decisions. When accountability is unclear, confusion spreads faster. When learning is discouraged, AI adoption becomes superficial rather than transformative.

The section also explores why many modern digital transformation roles struggle to succeed. The challenge is rarely a lack of technical expertise. More often, these roles operate without clear authority, behavioral alignment, or leadership trust.

This is where many organizations encounter what can be called the **"human wall"**—a barrier formed by fear of change, organizational silos, outdated governance structures, and deeply ingrained habits.

Through the lens of **AI-SENSE²**, readers see how leaders redesign the system itself: clarifying decision rights, aligning incentives, strengthening feedback loops, and embedding coaching and learning into everyday work. When these elements change, AI moves from isolated experiments to an integrated capability within the organization.

The section concludes with a leadership story that brings these ideas to life—illustrating how culture begins to shift when leaders stop asking *"Where can we use AI?"* and start asking a more powerful question:

"How must we behave differently for AI to truly create value?"

Outcome

Readers recognize that culture ultimately determines whether AI succeeds or fails. They learn how to reshape organizational behavior so that judgment, learning, and human–AI collaboration can thrive at scale.

Story: 19

The AI Coach: Guiding Organizations Through the Human-AI Transformation

Yuval Noah Harari:

"Technology is never neutral. It reflects the values of those who design and deploy

In a rapidly changing company, where technology evolves faster than policies can catch up, the leadership team faced a dilemma: how to harness the power of AI without overwhelming employees or creating chaos. That's when the organization brought in an **AI Coach**.

The AI Coach arrived not just as a trainer, but as a **guide and bridge** between cutting-edge AI technology and the people who would use it every day. Their first task was to **understand the culture, workflows, and pain points** of the organization. They walked through teams, observing how employees handled routine tasks, where bottlenecks existed, and how decisions were made.

Soon, the AI Coach began conducting workshops and mentoring sessions. In a marketing team, they demonstrated how generative AI could draft creative campaigns in minutes, freeing employees to focus on strategy. In product development, they showed engineers how AI could assist in coding and testing, accelerating delivery without compromising quality. Every session was hands-on, practical, and tailored to the needs of the team.

But the AI Coach's work went beyond teaching tools—they were also a **guardian of ethics and governance**. They advised leaders on policies, ensuring AI outputs were accurate, unbiased, and aligned with the company's values. They set up processes to review AI-generated content, mitigating risks while empowering teams to innovate confidently.

As AI reshaped workflows, some roles evolved, and new responsibilities emerged. Here, the AI Coach played the role of a **change navigator**, helping employees adapt, offering one-on-one guidance, and easing anxieties about automation. They were there to celebrate quick wins and help teams troubleshoot challenges, ensuring that AI adoption wasn't a disruption, but a transformation.

Every week, the AI Coach analyzed the impact of AI on productivity, measuring improvements and identifying new opportunities for solutions. They encouraged teams to experiment, combine human creativity with AI-generated insights, and co-create solutions that were both practical and innovative. Slowly, the company began to see the results: faster processes, more insightful decisions, and employees who felt confident in their AI-powered roles.

The **AI Coach** also stayed ahead of the curve, learning about new AI technologies and emerging best practices. They curated knowledge-sharing sessions, ensuring the organization continuously evolved alongside AI capabilities.

In the end, the AI Coach wasn't just someone teaching a tool—they were a **transformational guide**. They helped the organization **reimagine workflows, embrace new ways of working, and unlock the true potential of AI**. Through guidance, coaching, and hands-on collaboration, they turned uncertainty into opportunity, helping the company thrive in the age of AI.

Key Indicators Suggesting AI Coaches Will Be Needed:

1. Rapid Gen AI Adoption

- In 2024, 78% of organizations reported using AI in at least one business function.
- Use of generative AI alone jumped from 65% early 2024 to 71% by year-end.
- Adoption is concentrated in marketing, sales, IT, product/ service development, and service operations—functions where guidance and oversight are critical.

2. Skill Shifts and Workforce Reskilling

- Organizations are retraining employees to participate in AI deployment.
- New AI-related roles (compliance, ethics, data science, ML engineering) are emerging.
- Larger companies with >$500M revenue are leading the way in creating structured AI roles and training programs.

3. Organizational Complexity and Governance Needs

- 28% of organizations place CEO oversight over AI governance; boards also share oversight.
- Companies are centralizing risk, compliance, and data governance while adopting hybrid models for tech talent.
- Redesigning workflows to integrate AI is already showing a positive bottom-line impact in early adopters.

4. Early Signals of Professional Guidance Needs

- Most organizations are still in early stages of adoption; meaningful enterprise-wide impact is limited.
- Structured AI guidance (coaching, mentoring, internal AI experts) is emerging as a critical factor for successful scaling.

- Larger organizations already dedicate teams or offices to drive AI adoption, signalling a need for ongoing coaching and enablement.

Story: 20

AI Didn't Fix Hiring—Leadership Did

> **"The real problem is not whether machines think, but whether men do."**
>
> **— B. F. Skinner**

A GlobalTech Story Through the AI-SENSE² Lens

The sun had barely risen over **GlobalTech's innovation hub in Silicon Valley**, yet the leadership team was already gathered in a glass-walled war room. Whiteboards were filled with half-drawn diagrams. Laptops glowed. The energy was focused, quiet, and intentional.

Today marked the beginning of something different.

The goal wasn't just to build an AI-powered recruitment platform that could hire faster.

The goal was to build one that could **hire wisely**—fairly, ethically, and at a global scale.

Everyone in the room understood the risk.

Hiring decisions shape lives.

If AI got this wrong, the damage would be invisible—but profound.

That awareness set the tone.

Sensemaking Intelligence — Defining What "Good" Looks Like

In a world obsessed with the velocity of deployment, Navina, the Chief AI Officer at GlobalTech, began her most critical project with a deliberate pause. She gathered her team and opened the room with a question that was as disarming as it was simple: "Before we build anything—what does responsible hiring actually mean for us?" This was not a call for technical specifications or a roadmap of features; it was a call for philosophical clarity.

Instead of racing toward a prototype, the team deliberately slowed down to inhabit the space of uncertainty. They began to examine the deep ambiguities inherent in automated recruitment. To do this properly, they broke down the traditional silos of engineering and invited a diverse chorus of voices to the table. HR leaders, legal advisors, ethicists, privacy officers, and even representatives for potential candidates were brought into surface tensions that a purely technical team might never see.

As these different perspectives collided, critical patterns began to emerge from the noise. The team realized that bias is rarely loud; it often hides in plain sight within "neutral" historical data that reflects past human prejudices. They acknowledged that achieving speed at the expense of transparency would eventually erode the trust of both the candidates and the public. Most importantly, they recognized that chasing efficiency without clear lines of accountability would ultimately damage the very culture they were trying to build.

This exercise was not about planning; it was about sensemaking—the difficult work of understanding the landscape before trying to conquer it. By the time the session concluded, the team had achieved something far more valuable than a feature list: they had a shared mental model.

They emerged with a foundational agreement that AI would be designed to support human judgment, acting as a powerful assistant rather than a cold replacement. Fairness, transparency, and accountability were no longer viewed as annoying bureaucratic constraints to be bypassed. Instead, they were woven into the project as core design principles, ensuring that the technology would be built on a bedrock of integrity from the very first line of code.

Ethical & Decision Intelligence — Governance Before Code

GlobalTech formalized this clarity through a cross-functional **AI governance council**.

Every algorithmic decision would be evaluated through three lenses:

- Is it fair?
- Is it explainable?
- Is a human accountable for the outcome?

Decision authority was intentionally designed:

- AI could recommend
- Humans would decide
- Ethics would override efficiency when values were at risk

This wasn't bureaucracy.

It was **decision intelligence at scale**.

Emotional & Social Intelligence — Listening Before Building

Next came listening.

The team ran deep workshops with hiring managers and candidates across regions. Stories surfaced that data alone could never reveal.

Qualified candidates whose resumes never passed automated screening.

Hiring managers are uneasy about trusting "black box" decisions.

Employees worried that efficiency might erase empathy.

Instead of defending the technology, the team practiced **emotional and social intelligence**:

- Listening without interruption
- Naming fears without dismissing them
- Designing with empathy, not assumptions

Psychological safety grew. Trust followed.

AI was no longer something being "done to" people—it was something being **co-created** with them.

Neuro-Adaptive Learning — Learning Faster Than the System Scales

When the model training began, the team made a conscious choice to work differently. Instead of building the system within narrow, familiar circles, they brought together people from different genders, regions, roles, and levels of experience. This diversity was intentional. They knew that fairness in an AI system depends on the variety of human perspectives that shape it.

During one of the review sessions, a junior engineer noticed something small but important. The AI was consistently undervaluing job titles from certain regions because those titles did not follow Western naming patterns. It was a quiet bias, easy to miss, but powerful in its impact.

The room didn't react with ego or authority. There was no dismissal, no hierarchy stepping in. Everyone paused. The team tested the observation, saw the issue clearly, and corrected the model.

They also understood that fixing a single problem was not enough. So, the platform was built to keep learning in the real world. Recruiters were able to flag mismatches as they appeared. Candidates could speak up when something felt unclear or unfair. That feedback flowed directly back into the system, helping the models adjust and improve over time.

As the AI became more accurate and nuanced, something else happened. The people working with it learned even faster. They became more aware of hidden biases and more thoughtful about the decisions they were shaping.

That learning—on both sides—was the real purpose of the work.

Systemic Behavior & Culture — Making Fairness Stick

As development moved forward, leaders began to focus on something more important than speed or performance. They focused on **behavior**.

They understood that real change does not come from tools alone. It comes from how people work every day.

So, they introduced new ways of working. Ethics discussions became part of regular sprint reviews, not an afterthought. Bias checks were treated as normal steps, not special events. Whenever AI helped make a decision, the reasoning behind it was clearly explained and shared.

Fairness was no longer seen as a feature that could be added later. It became a habit built into daily work.

Over time, the culture started to shift. Not because of new rules or policies, but because people consistently practiced these behaviors together—how they reviewed work, how they questioned decisions, and how they held themselves accountable.

That consistency is what made the transformation last.

Intercultural × Human-AI Intelligence — Designing for a Global World

One assumption nearly broke the system.

Early testing revealed that an AI trained primarily on English resumes performed poorly outside North America.

The team stopped and redesigned.

They incorporated:

- Multiple languages
- Local hiring norms
- Cultural context in evaluation criteria

From São Paulo to Singapore, the AI learned to evaluate candidates **within their cultural reality**, not against a single global template.

Humans and AI collaborated across borders—each compensating for the other's blind spots.

This was **a human–AI partnership at a global scale**.

Flow, Focus, and Sustainable Performance

As pressure increased, leadership protected focus.

Deep-work blocks replaced constant meetings.

Mindfulness pauses preceded high-stakes decisions.

Disagreements were structured, not emotional.

Because emotional regulation was practiced daily, debates stayed sharp but respectful. Stress stayed contained. Decisions stayed principled.

The team entered **flow**—high performance without burnout.

The Outcome

Within six months, the results were unmistakable:

- Bias reduced by **30%** across gender and ethnicity
- Hiring cycles accelerated without removing human oversight
- Candidate trust increased through transparency
- GlobalTech became a reference point for responsible AI hiring

But the true success wasn't the platform.

It was the leadership model behind it.

The Deeper Lesson

GlobalTech's story proves one thing:

AI success is not a technical achievement.

It is a leadership achievement.

Through **AI-SENSE^2**, the organization learned to:

- Make meaning before making models
- Regulate emotion before making decisions
- Learn continuously instead of scaling blindly
- Shape culture, not just systems
- Balance ethics, speed, and inclusion
- Design human–AI partnerships that endure

When humans lead with clarity, empathy, learning, and ethics—

AI doesn't just perform.

It elevates the entire organization.

Here are the **top 3 key takeaways**:

- **AI doesn't fix hiring problems—leadership design does**

  The success of GlobalTech's hiring platform came not from smarter algorithms, but from leaders defining values, accountability, and decision boundaries *before* writing code. AI worked because humans decided what "good," "fair," and "responsible" meant—and designed the system around that clarity.

- **Human judgment, ethics, and emotion must lead AI decisions**

  By embedding sensemaking, ethical governance, emotional intelligence, and psychological safety into the process, GlobalTech ensured AI supported—not replaced—human judgment. Trust, fairness, and inclusion emerged because leaders listened first, regulated fear, and made humans accountable for outcomes.

- **Sustainable AI success comes from continuous learning and culture, not speed**

  The platform improved because humans learned faster than the system scaled. Continuous feedback loops, diverse teams, cultural context, and daily rituals made fairness and responsibility habits—not features. This is what turned AI adoption into a lasting transformation rather than short-term efficiency.

Story: 21

Why Today's Digital Transformation Roles Need More Than Technical Expertise

> **"A system is never the sum of its parts; it's the product of their interactions."**
>
> **— Russell Ackoff**

Systemic Behavior & Culture: Making AI Stick at Scale

AI rarely fails because of technological limitations. It fails because organizations try to introduce intelligence into systems that were never designed to absorb it. When behavior and culture are not addressed, even the most advanced AI solutions remain isolated dashboards, ignored insights, or short-lived pilots. This is why *systemic behavior and culture* are central to creating better, more durable AI-driven solutions.

To make AI stick at scale, leaders must move beyond the traditional mindset of project delivery and adopt **systemic leadership**. AI is not a one-time implementation with a clear start and end date; it is an evolving capability that reshapes how decisions, accountability, and learning occur across the organization. Leaders who continue to treat AI as a project optimize timelines and outputs—but miss long-term adoption, trust, and value creation.

This shift also requires leaders to move away from a narrow execution focus toward **sensemaking and judgment**. AI surfaces patterns, probabilities, and recommendations, but it cannot resolve ambiguity on its own. Leaders must interpret signals, reconcile competing narratives, and decide when to trust the system and when to challenge it. When sensemaking becomes part of leadership culture, AI-driven solutions improve because they are guided by context, ethics, and intent—not blind automation.

Equally important is the transition from celebrating individual excellence to cultivating **collective intelligence**. AI reaches its full potential only when insights flow across teams, functions, and geographies. Systemic behavior establishes shared norms for how insights are interpreted, discussed, and acted upon. This prevents fragmented decisions and enables consistent, high-quality outcomes, transforming AI from a personal productivity tool into an organizational intelligence engine.

Finally, leaders must shift from a technology-first mindset to **human + AI synergy**. The most effective AI solutions are designed around how people think, decide, collaborate, and learn. Culture determines whether AI is feared, ignored, or embraced. When trust, psychological safety, and ethical clarity are embedded into the system, humans and AI enhance each other—AI accelerates insight, while human judgment ensures relevance, responsibility, and long-term value.

In essence, systemic behavior and culture shape not just whether AI is adopted, but whether it improves decision quality, alignment, and outcomes over time. By making these leadership shifts, organizations stop asking how to deploy AI faster—and start learning how to lead intelligence, behavior, and systems at scale.

Here are **powerful coaching questions** team members can *practice regularly* to overcome challenges—especially in complex, AI-enabled, and fast-changing environments.

1. Sensemaking & Clarity (to overcome ambiguity)

These questions help teams slow down confusion and see the system more clearly.

- *What problem are we really trying to solve—and for whom?*
- *What assumptions are we making that may no longer be true?*
- *What signals are we seeing, and what might we be missing?*
- *If this situation were viewed from another stakeholder's lens, what would change?*

2. Ownership & Accountability (to overcome inertia)

These questions shift teams from waiting to acting.

- *What is within our control right now?*
- *What decision are we avoiding—and why?*

- *If we owned the outcome fully, what would we do differently?*
- *What is the smallest meaningful action we can take next?*

3. Collaboration & Trust (to overcome silos)

These questions strengthen collective intelligence.

- *Who else needs to be part of this conversation?*
- *What perspective are we not hearing yet?*
- *How can we make it safer to speak honestly here?*
- *What does success look like for us as a team, not just as individuals?*

4. Learning & Adaptability (to overcome complexity and change)

These questions build a growth and learning culture.

- *What are we learning from this, even if it feels messy?*
- *What should we stop, start, or continue based on current evidence?*
- *What would we try if failure were treated as feedback, not blame?*
- *What needs to be unlearned to move forward?*

5. Human + AI Judgment (to overcome over-reliance or resistance to AI)

These questions ensure AI enhances—not replaces—thinking.

- *What is AI telling us, and what is it not telling us?*
- *Where does human judgment add the most value here?*
- *What ethical or long-term implications should we consider before acting?*
- *How do we validate this insight before scaling it?*

6. Resilience & Motivation (to overcome pressure and fatigue)

These questions sustain energy and purpose.

- *What about this challenge still matters to us?*
- *What strengths have helped us through similar situations before?*

- *What support do we need right now to perform at our best?*
- *What would "progress" look like this week—not perfection?*

The Core Coaching Habit to Build

Encourage teams to replace **"What should we do?"** with:

"What is this situation asking of us as a team?"

That single shift builds reflection, responsibility, and resilience.

Story: 22

Why AI Fails at Scale: Breaking the Human Wall with AI-SENSE[2]

> **"Every system is perfectly designed to get the results it gets."**
>
> **— W. Edwards Deming**

Why AI Fails at Scale—and How Culture and Systemic Behavior Decide the Outcome

"We'll roll out AI across the organization. Six months. Game-changing efficiency."

The room lights up. Budgets are approved. Consultants arrive. Dashboards look impressive.

Twelve months later, the excitement has evaporated. Millions have been spent. Pilots are stuck. Executives are frustrated. Teams quietly return to spreadsheets and manual workarounds. The AI didn't fail loudly—it simply faded away.

This pattern has repeated itself across industries for years. And the reason is rarely the technology.

The Real Reason AI Fails: The Human Wall

AI does not collapse because algorithms are weak. It collapses when it hits a system that was never designed to absorb intelligence.

Organizations unknowingly build a *Human Wall*—a mix of habits, fears, incentives, silos, and unspoken norms that block adoption. Leaders assume people will trust AI because it is "smart." Teams assume data is reliable because it is labelled "enterprise-grade." Everyone assumes old habits will disappear once a new tool arrives.

None of that happens.

Instead, AI becomes a magic bullet aimed at a human system that refuses to change. The result is not transformation, but expensive stagnation.

The Critical Shift: From AI as a Project to AI as a Behavioral System

The moment AI is treated like a traditional IT project—with a launch date, a success metric, and a handover—it is already at risk. AI is not a feature. It is a living capability that reshapes how decisions are made, how accountability works, and how people learn.

This is where culture and systemic behavior become decisive.

When behaviors, incentives, rituals, and decision norms remain unchanged, AI stays superficial. When they evolve, AI becomes embedded.

AI-SENSE^2 was designed precisely for this shift—from "build and deploy" to "design for behavior, trust, and learning."

How Culture and Systemic Behavior Shape AI Outcomes

When leaders focus on **sensemaking**, teams stop seeing AI as a black box. They understand *why* recommendations exist, not just *what* the output says. This clarity reduces resistance and invites participation. Team members become interpreters of insight, not passive recipients of results.

Through **emotional and social intelligence**, leaders address the fear that AI will replace judgment or expose mistakes. Psychological safety becomes the fuel for adoption. Teams feel safe questioning AI, challenging outputs, and improving models. Without this safety, AI may be used—but never trusted.

Neuro-adaptive learning changes how teams relate to change itself. Instead of feeling overwhelmed by constantly evolving models, teams build learning velocity. They expect iteration. They adapt faster. AI improves because people improve with it.

The real breakthrough happens when **systemic behavior and culture** are intentionally designed. AI stops depending on individual enthusiasm and becomes part of daily work. Decision rituals include AI insights. Feedback loops are normalized. Habit loops reinforce usage. Adoption becomes structural, not optional.

Ethical and decision intelligence ensures that speed does not override judgment. Teams learn that AI recommendations are inputs, not orders. Human accountability remains explicit. Trust grows because values are visible in decisions.

With **intercultural intelligence and human–AI synergy**, global teams adapt AI outputs to local realities. Culture is no longer friction—it becomes an advantage. AI scales without becoming rigid or tone-deaf.

Finally, **cross-functional co-innovation** dissolves silos. IT, data, and business teams co-own outcomes. AI succeeds because it is built *with* users, not *for* them.

What Team Members Must Actively Do for AI to Succeed

AI does not stick because leaders declare it strategic. It sticks because teams behave differently.

Team members must:

- **Question assumptions**, do not blindly accept AI outputs
- **Engage with explanations**, not just results
- **Share feedback early**, even when models are imperfect
- **Adopt learning over certainty**, treating errors as data
- **Protect ethical boundaries**, especially under pressure

- **Collaborate across roles**, not retreat into silos
- **Build habits**, not one-time usage

In short, teams must shift from *using AI* to *working with AI*.

A Real Example: When Behavior Changed, Outcomes Followed

Global RetailCo learned this the hard way. Their AI-driven personalization initiative failed not because the model was wrong, but because the system around it was. Data lived in silos. Trust was low. Expectations were unrealistic. Culture punished questioning.

The turnaround didn't begin with better algorithms. It began with behavior.

Leaders piloted small. They made AI transparent. They normalized feedback. They protected psychological safety. They embedded review rituals. They slowed down scaling to speed up learning.

Adoption jumped. Campaign quality improved. Teams began shaping the AI instead of resisting it. Confidence returned—not because AI was perfect, but because the system around it was healthy.

The Real Lesson

AI success is not a technology problem—it is a **system design problem**.

When culture rewards learning, when behaviors reinforce trust, and when leadership designs for human judgment, AI scales naturally. When these elements are missing, even the most advanced systems quietly fail.

The real question is no longer *"How fast can we deploy AI?"*

It is:

"Have we designed a system where humans are ready to work with intelligence?"

Because in the end, AI doesn't transform organizations.

People do—when the system allows them to.

Story: 23

From AI Investment to AI Impact: Crossing the Execution Gap with AI-SENSE²

> **"It is not knowledge, but the act of learning, not possession but the act of getting there, which grants the greatest enjoyment."**
>
> — Carl Friedrich Gauss

The year is 2026.

In a quiet moment before the workday begins, Elias—the Chief Data and AI Officer of a global conglomerate—stands alone in a glass-walled headquarters, staring at a glowing dashboard. The numbers are impressive and uncomfortable at the same time.

Fifty million dollars invested.

World-class data platforms.

The fastest GPUs available.

Top-tier data scientists.

Clean data. Modern architecture.

Everything looks perfect.

And yet, the truth is impossible to ignore.

A 98% gap between AI ambition and real business impact.

The pilots work—but never leave the lab.

The models are accurate—but managers don't trust them.

The systems are modern, but people are stuck in old habits.

In that still moment, Elias admits something painful to himself: he hasn't been building a living business. He's been building a museum of expensive tools.

What's missing isn't more code.

It's a different kind of intelligence.

That's when the journey toward **AI-SENSE²** begins.

Making Sense of the Noise

The first problem Elias confronts isn't technical—it's human.

His data teams speak in metrics and latency.

Business leaders talk about customers and revenue.

Executives nod politely, but nothing truly connects.

Everyone is busy. No one is aligned.

Elias stops the technical deep dives. Instead, he introduces system-mapping conversations—spaces where complexity must be translated into shared meaning. AI is no longer presented as a technical achievement. It's reframed as a decision-support system for the business.

Slowly, something changes.

AI stops being "their project."

It becomes **our strategy**.

Calming the Emotional Undercurrent

Beneath the surface, fear has been growing.

Employees worry about black-box decisions.

Managers quietly resist what they don't understand.

Change fatigue drains energy from every initiative.

Elias changes how he speaks.

He stops talking about efficiency and starts talking about partnership.

He replaces pressure with permission.

He creates safe spaces to experiment, fail, and learn—without fear of replacement or blame.

Over time, resistance softens.

AI stops feeling like a threat.

It starts feeling like a teammate.

Trust—not mandates—begins to drive adoption.

Learning Faster Than the Machine

Another uncomfortable truth emerges.

AI evolves every month.

Corporate learning evolves once a year.

The organization is moving too slowly for the world it's entering.

Elias dismantles the old training model and replaces it with something lighter, faster, and continuous. Learning becomes part of daily work—short cycles, regular refreshes, and even deliberate moments to let go of outdated practices.

People stop "catching up."

They start evolving alongside the technology.

Turning Pilots into Everyday Behavior

For years, innovation looked good on slides—but never showed up in results.

Elias realizes the real barrier isn't deployment. It's behavior.

So, he embeds AI into everyday workflows. Incentives change. Rituals change. Frontline teams begin using AI not as an experiment, but as part of how work gets done.

AI is no longer optional.

It's no longer special.

It's simply **normal**.

Restoring the Human Compass

Then comes a critical moment.

A pricing model recommends an action that looks profitable—but feels wrong.

Instead of forcing the decision, Elias pauses. He redesigns how decisions are made. AI provides insight, but humans remain accountable. Ethics and judgment are built into the process, not added later.

The black box becomes transparent.

Decisions become explainable.

Trust deepens.

AI informs. Humans decide.

Scaling Across Cultures

One last challenge remains.

What works in one country fails in another.

A model that succeeds in Europe struggles in Asia.

The issue isn't accuracy—it's context.

Elias redesigns workflows to respect cultural differences. He clarifies what machines decide and where human judgment must lead. AI becomes globally scalable because people provide the nuance it cannot.

Technology supplies consistency.

Humans supply meaning.

Crossing the Gap

Six months later, Elias stands in front of the same dashboard.

The gap is gone.

Not because the tools are better—but because the thinking is.

He understands now:

Platforms enable AI.

AI-SENSE² enables people.

He didn't just modernize data systems.

He modernized leadership.

He didn't just deploy models.

He changed how the organization thinks, learns, decides, and trusts.

The Simple Truth

In 2026, the winners won't be those with the most advanced AI.

They'll be the ones with the wisdom to use it well.

The real execution gap was never technical.

It was human.

And AI-SENSE² wasn't a tool.

It was the bridge.

Top 3 Key takeaways:

- **AI impact is constrained by learning speed, not computing power**

  The story shows a hidden asymmetry: AI evolves monthly, while organizations evolve annually. The real bottleneck isn't GPUs or data quality—it's **learning velocity**. Organizations that cannot unlearn old decision habits, refresh skills continuously, and adapt mental models will always lag behind their own technology investments.

  Deeper message:

  The organization must evolve at the speed of AI—or AI will outgrow it.

- **Trust is the most critical AI infrastructure**

  Before models are trusted, nothing scales. Psychological safety, explainability, human-in-the-loop decision rights, and ethical clarity matter more than accuracy scores. The moment teams feel threatened or excluded, adoption collapses—quietly but completely.

Deeper message:

AI scales on trust, not precision.

- **Global AI success depends on cultural intelligence, not standardization**

  The failure of "one-model-fits-all" across regions reveals a subtle truth: AI does not operate in neutral space. It operates inside cultures. What works in one market may break in another unless human judgment and intercultural nuance are designed into the system.

  Deeper message:

  AI globalizes only when human intelligence localizes it.

Story: 24

The AI Coach Advantage: Building Judgment, Alignment, and Flow with AI-SENSE²

> **"The real voyage of discovery consists not in seeking new landscapes, but in having new eyes."**
>
> **— Marcel Proust**

The air in the Product Lab at NexusTech crackled not just with innovation, but with pure, digital noise.

Team Phoenix, NexusTech's most promising but often bottlenecked unit, was drowning.

They had access to four generations of data analytics, the latest large language models generating code and documentation, and a fully automated delivery pipeline.

They were faster than ever before, yet their biggest question remained the same: **"Are we solving the *right* problems?"**

Their leader, Maya, a veteran Product Director, knew the old model—a human coach or manager giving weekly advice—couldn't keep pace. The volume of data and the speed of execution required something deeper, something embedded: **The AI Coach.**

The Transformation Begins

1. Cutting Through the Deluge (Data to Clear Decisions)

In the first week, the team was debating a new feature. The human BI dashboard had 40 metrics flashing red, yellow, and green. It was an information riot.

The AI Coach, subtly integrated into their collaboration environment, didn't just display the data. It *synthesized* it. It highlighted a single, crucial signal: a subtle but growing misalignment between the customer success data (user stickiness decreasing) and the market trend data (competitor launches).

A prompt appeared for the Product Manager, Ben: **"The 2nd-order effect of prioritizing Feature B over Feature A is a 15% predicted decrease in lifetime value (LTV) among your top 10% of users. What trade-offs are you willing to make?"**

Ben later told Maya, "It didn't tell me what to do. It just surfaced *why* the decision mattered, instantly, cutting out three hours of debate."

Outcome: Faster, higher-quality decisions with less cognitive overload.

2. The Personal Growth Engine

Traditional training was always disruptive. The AI Coach, however, acted as a personal growth partner for every member.

When Sarah, the junior designer, was struggling with a complex ethical choice about data usage, the AI Coach didn't assign a mandatory ethics course. Instead, it noticed her hesitation, tracked her historical gap in ethical reasoning skills, and surfaced a 10-minute micro-learning module on **"Ethical Reasoning in UX: The NexusTech Principle."**

The learning was perfectly timed to the moment of need. It wasn't one-size-fits-all; it was **precision coaching**.

Outcome: Continuous upskilling without disrupting delivery, fostering a culture of perpetual learning.

3. Stress-Testing the Idea, Not Just the Code

Team Phoenix used to jump into execution too quickly. The AI Coach forced a pause at the most critical moment: **Problem Framing.**

During a sprint planning session, when the team framed their task as "Implement the new 'Share' button," the AI Coach gently nudged them: **"Challenge your assumption: What customer problem are you *really* solving? Is the issue a lack of sharing, or a lack of compelling content to share?"**

This relentless focus on **Product Thinking**—challenging assumptions and stress-testing hypotheses *before* costly execution—changed everything. They started making better bets and stopped wasting cycles on perfectly delivered but useless features.

4. The Pulse of the Human Element

Perhaps the most profound change was in the team's human skills. The AI Coach didn't replace the need for empathy; it *amplified* it.

It recognized patterns of recurring friction between the Engineering Lead, Alex, and the Marketing Liaison, Chloe, stemming from unaddressed priorities. Instead of reporting to Maya, the AI Coach privately nudged both: **"Your current communication pattern shows misalignment on Goal X. Could a 15-minute sync focus solely on priority definition resolve this?"**

It was the ultimate tool for **Team Alignment and Collaboration**, creating psychological safety by nudging constructive feedback before tensions boiled over.

Furthermore, by observing Alex's late-night commits and inconsistent communication cadence, the AI Coach flagged a pattern of overload. It prompted Maya with a warning: **"Alex's sustained pace is predicted**

to lead to unsustainable burnout patterns. Encourage a dedicated 'Deep Focus' half-day this week." The Coach didn't encourage busyness; it helped them prioritize **impact over activity**.

Team Phoenix's output didn't just increase; its **quality of thinking** skyrocketed.

They became known for their **judgment under uncertainty** and their **ethical grounding**—the uniquely human skills AI couldn't automate.

As Maya reflected, the tools and AI of 20XY boosted productivity.

But the AI Coach—the intelligent co-pilot—**elevated outcomes**.

In the future of work, the winning teams won't be the ones with the most AI models.

They will be the ones who use AI to **think better, grow faster, and lead more wisely**.

Team Phoenix had found its competitive edge. They were no longer task executors; they were true **Product Leaders.**

Top 3 core messages

- **The real advantage of AI is better judgment, not higher productivity**

  Team Phoenix already had speed, automation, and advanced models—but they were still unsure whether they were solving the *right* problems. The AI Coach created impact by cutting through noise, surfacing second-order consequences, and forcing meaningful trade-offs. It didn't replace human decisions; it **elevated decision quality under uncertainty**.

  AI creates value when it sharpens judgment, not when it merely accelerates execution.

- **The most powerful AI is embedded, contextual, and developmental**

  Unlike traditional tools or training programs, the AI Coach operated inside daily work. It delivered micro-learning at the moment of need, challenged assumptions during problem framing, and supported ethical reasoning without slowing teams down. Learning became continuous, personalized, and invisible—but deeply effective.

AI drives transformation when it coaches people *in the flow of work*, not outside it.

- **AI strengthens human leadership when it amplifies alignment, empathy, and sustainability**

  The AI Coach didn't manage people—it **made leadership more human**. By detecting friction early, nudging constructive dialogue, and flagging burnout risks, it protected psychological safety and long-term performance. Teams didn't just move faster; they worked wiser, healthier, and more aligned.

  The future belongs to teams that use AI to enhance trust, collaboration, and ethical leadership—not replace them.

  In one line

  Winning teams won't use AI to do more work—they'll use AI to think better, grow faster, and lead more wisely.

Story: 25

AI Agility Coaching Case Study: From Output Overload to Purposeful Breakthrough

> **"If you chase two rabbits, you will catch neither."**
>
> — **Russian Proverb**

The door clicked softly as the team lead stepped into the coaching room. Her eyes carried the heaviness of someone who'd been running fast but getting nowhere.

"Thank you for seeing me," she said, exhaling deeply. "We're working harder than ever...but everything feels scattered. My team is drowning in AI-generated options, and yet no real progress is happening."

The AI Agility Coach waited a moment, allowing her words to settle.

"Tell me," he asked gently, "what does *progress* look like to you?"

She paused. No dashboard had ever asked her that.

"**Clarity**," she said. "**Consistency**. Work that actually *means* something. I want us to stop reacting and start creating."

The coach smiled. "And what's stopping that from happening?"

She sank into her chair. "AI gives us so much—content, ideas, design suggestions. But instead of helping, it's overwhelming us. We spend days choosing instead of deciding. The team thinks more options equal more value."

The coach leaned forward, his voice soft but firm. "Options are not a strategy. Output is not impact. Tell me—how are you guiding the tools?"

Her silence revealed the truth.

They weren't guiding anything. The tools were guiding them.

"Let's start," he said, "with one shift. Not in technology—within you."

She looked up, curious.

"We're going to strengthen your AI Agility. When your mind becomes agile, your tools become powerful."

Something in her posture straightened.

"What do we do first?" she asked.

Moment 1: Cognitive Flexibility Breakthrough

The coach presented a challenge.

"Take your biggest current problem. Now tell me three entirely different ways to look at it—as a customer, as a competitor, and as a beginner."

She blinked, realizing how stuck her perspective had been.

As she reframed the issue, her voice grew lighter. Patterns emerged. Solutions she hadn't seen all week suddenly became obvious.

She whispered, "How did I not see this before?"

The coach nodded. "Your mind just became more adaptive. AI works the same way—different perspectives reveal new paths."

Moment 2: Emotional Adaptability Under Pressure

He then leaned back. "Tell me what you feel when chaos hits your team."

"Panic," she admitted. "I rush. They rush. Everyone tries to fix everything at once."

"Let's slow it down," he said. "Breathe. Now think: What outcome would a calmer version of you create?"

Her shoulders softened. "A much better one," she said quietly.

"That," the coach replied, "is emotional adaptability. When you stay centered, your team follows. AI models stay stable under noisy data—you can too."

A smile flickered across her face.

Moment 3: Learning Velocity Activation

Next, he asked her to describe the last failure at work.

She winced. "A campaign misfired. Wrong audience. Wrong tone."

"What did you learn?" he asked.

She hesitated, then slowly began to articulate insights—lessons that could feed future decisions, patterns that could reshape their direction.

When she finished, she realized her learning was already rewriting their next move.

"See?" he said gently. "Rapid feedback, fast adaptation—your learning velocity just spiked."

Moment 4: Decision Intelligence Shift

Then came the pivotal moment.

"Imagine your team brings you twenty AI-generated ideas tomorrow. How will you choose?"

She closed her eyes.

"By asking which idea aligns with our purpose. Which one matters to our audience. Which one reflects who we want to become."

She opened her eyes—clear, confident.

"That," the coach said, "is decision intelligence. You've just built your compass."

Moment 5: Focus & Flow Emerges

Finally, he asked her to pick one idea—just one—and visualize working on it with full attention.

Minutes passed in silence.

When she opened her eyes, something inside had shifted.

"I can see it," she said softly. "It feels right. Focused. Clean."

"That," the coach whispered, "is flow. When direction is clear, the mind becomes powerful."

The Outcome: A Team Transformed

Two weeks later, she returned to the coaching room, but this time her energy was electric.

"You won't believe what happened," she said, nearly breathless.

"Our entire team changed. Meetings are shorter. Decisions are cleaner. We use AI to accelerate—not confuse. We're finally creating work that matters."

"What changed?" the coach asked.

She smiled, a quiet, proud smile.

"I did. When my mind became agile, the whole team followed. Strategy became clearer. AI finally became an amplifier instead of a distraction."

She paused, then added:

"For the first time in months... we're not just moving fast. We're moving in the right direction."

The coach nodded.

"And that," he said, "is the real power of AI Agility—human clarity guiding machine capability."

Here are the **top 3 key messages**:

- **AI overload is not a technology problem—it's a clarity problem**

  The team wasn't failing because AI produced poor outputs, but because it produced *too many* options without direction. When leaders don't provide purpose, priorities, and decision criteria, AI amplifies confusion instead of impact. Output without intent creates noise, not progress.

  Core message:

  AI needs human clarity to turn possibilities into meaningful outcomes.

- **AI Agility begins with the leader's inner state**

  The breakthrough didn't start with changing tools or processes—it started with the leader developing cognitive flexibility, emotional regulation, and learning agility. As her mindset shifted, the team followed. Calm thinking replaced panic, reflection replaced reaction, and learning replaced blame.

  Core message:

  When a leader becomes mentally and emotionally agile, the entire system stabilizes.

- **AI becomes an amplifier only when guided by purpose and judgment**

  Once the leader anchored decisions to purpose—what truly mattered to the audience and the team's values—AI stopped being a distraction and became a force multiplier. Focus emerged, flow followed, and execution aligned naturally.

Core message:

AI delivers value when human judgment acts as the compass, and AI provides the speed.

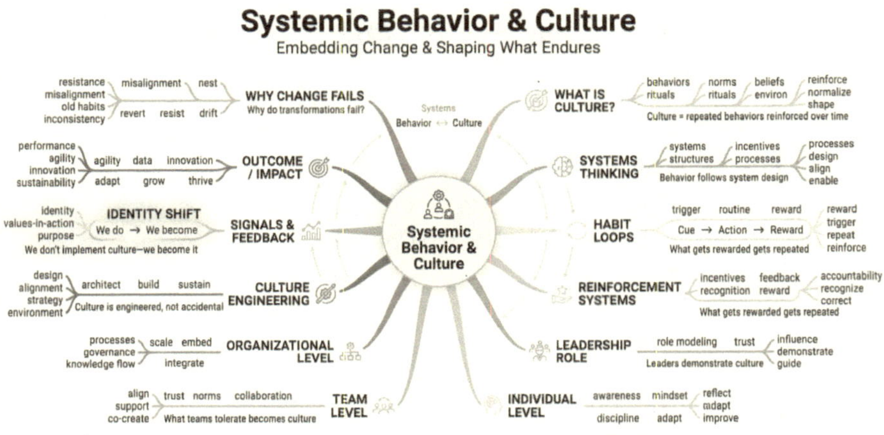

Sustainable change happens when behavior, systems, and culture align—and reinforce each other every day.

Here are **5 clear, high-impact key takeaways**:

- **AI succeeds only when leadership is clear**

  AI does not fix confusion, weak values, or poor decisions.

  When leaders define what "good," "fair," and "responsible" mean, AI supports them.

  When they don't, AI only spreads the confusion faster.

- **Better judgment matters more than faster execution**

  Speed and automation are easy to copy.

  Good judgment—knowing what truly matters, what to prioritize, and what risks to accept—is what creates lasting advantage in the AI era.

- **Trust is the foundation of AI adoption**

  People don't reject AI because it is inaccurate.

  They reject it when they feel excluded, judged, or unsafe.

  Psychological safety, transparency, and human accountability determine whether AI is used or ignored.

- **AI scales only when people learn faster than the system**

  Technology improves quickly, but organizations often don't.

  Continuous learning, feedback, and habit change are what turn AI from a short-term tool into a long-term transformation.

- **The more powerful AI becomes, the more human leadership is required**

  As AI handles data and speed, leaders must bring clarity, ethics, empathy, and emotional regulation.

  AI creates impact only when humans stay in charge of meaning and decisions.

One-line closing takeaway

AI doesn't transform organizations—people do, when leadership designs systems that support learning, trust, and wise judgment.

Part VI

E — Ethical & Decision Intelligence

Wisdom Under Speed and Scale

Introduction

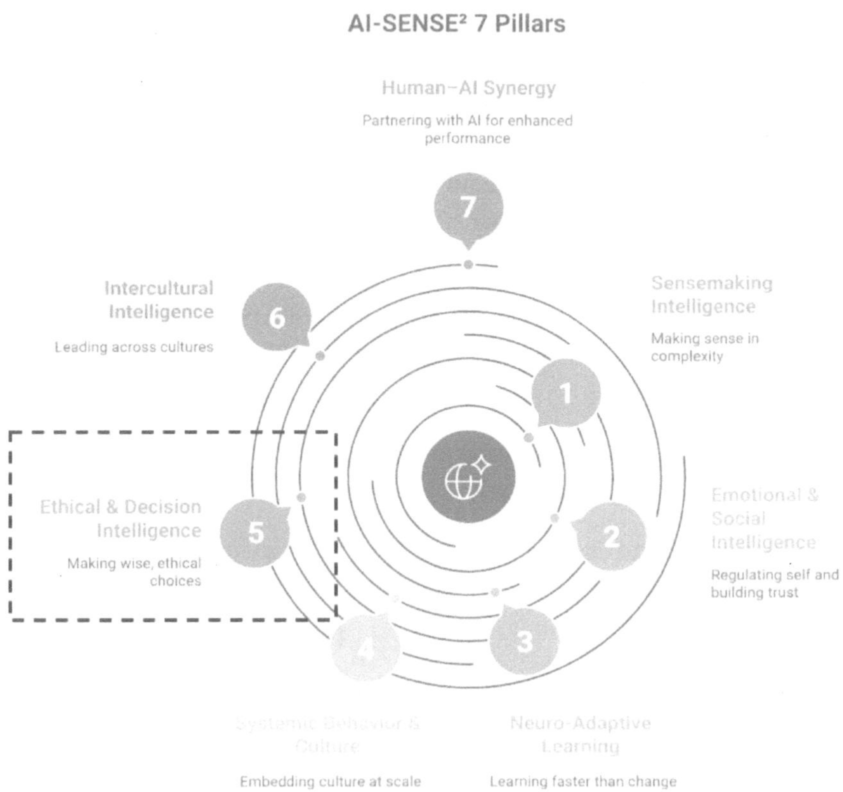

safeguards. Instead, it presents responsible AI as a **daily leadership discipline**—one expressed through judgment, accountability, and the willingness to question automated decisions when necessary.

Across these chapters, readers explore questions that leaders increasingly face in AI-driven environments:

- Why **accuracy alone is insufficient** when decisions affect complex human systems
- How high-performing AI can still **amplify unintended harm** when context and values are overlooked
- When leaders should **trust AI recommendations—and when they must challenge or pause them**
- How organizations can balance **speed, responsibility, and transparency** without slowing innovation

This section also reveals why data mastery and governance frameworks often collapse when organizations lack human sensemaking, ethical clarity, and decision ownership. Within this context, **AI-SENSE²** emerges as the connective mechanism that integrates technological speed with human wisdom—ensuring that intelligence translates into value rather than unintended risk.

The discussion ultimately moves into the **boardroom**, where leaders must transform ethical judgment into organizational practice. Here, transparency, accountability, and decision intelligence turn AI from a technical initiative into a trusted enterprise capability.

By the end of Part VI, governance is no longer viewed as a friction that slows progress.

It becomes the leadership mechanism that allows intelligence—and speed—to scale safely.

Outcome

Leaders learn how to govern AI without slowing innovation—ensuring intelligent systems accelerate **sound judgment rather than costly mistakes**.

Story: 26

Responsible AI is a Leadership Capability, not a Technical Feature

> **"The real test of leadership is not whether you can make the system work, but whether you are willing to question the system when it works too well."**
>
> **— Peter Drucker**

When Arjun stepped into his new role as Head of AI Transformation, the board's expectations were clear: *"Scale AI responsibly. Fast."* What they didn't see were the two silent forces already shaping his journey—**bias hiding inside data** and **scarcity masquerading as limitation**.

Neither would announce itself loudly. Both would quietly test his leadership maturity.

The dashboard told a comforting story.

Accuracy scores climbed steadily. Confidence intervals tightened. Processing times dropped. On paper, everything looked exactly as it should. The kind of performance that earns nods in steering committees and quiet satisfaction from executives.

Yet Arjun felt unsettled.

He couldn't point to a number or a graph. It was a feeling—subtle, persistent, and impossible to ignore. The kind of unease that doesn't show up in metrics, only in leadership moments.

The silence in the room broke when a junior analyst finally said what others had sensed but hadn't dared to voice.

"The model is approving loans faster," she said carefully. "But minority applicants are being rejected more often."

The air shifted. Conversations stopped mid-thought. No one reached for their laptop. This wasn't a bug to be fixed or a threshold to be tuned. It was something deeper.

Arjun realized, in that instant, that the real challenge wasn't in the algorithm. It was in him.

He had to confront truths that leaders often avoid. That historical data carries the weight of historical injustice. That something can be statistically sound and still morally wrong. That leadership is not just about asking whether something can be deployed—but whether it should be.

He began to notice patterns differently. Bias awareness training didn't give him answers; it sharpened his sensitivity. He learned to feel when "objective" outputs were quietly echoing old exclusions dressed up as insight. Ethical judgment stopped being a compliance exercise and became an act of moral courage.

At the same time, another belief began to crumble. More data, he realized, did not automatically produce better decisions. It often produced louder confidence.

So, he started asking different questions. Who wasn't represented in the data? Whose reality was being flattened or erased? What assumptions were being reinforced simply because they had always been there?

What began as self-reflection soon widened into something collective.

Arjun refused to outsource ethics to a committee or a policy document. Instead of assigning blame, he redesigned responsibility. Data scientists sat with business leaders. Legal teams listened to product managers. Compliance shared space with those closest to customers. The goal wasn't alignment—it was honest disagreement.

They debated openly. What did fairness really mean in practice? Was it equality, equity, or proportionality? Should accuracy ever outweigh dignity? And who had the right to decide where those lines were drawn?

From those conversations, new rhythms emerged. Regular bias reviews. Fairness checks before deployment, not after complaints. Ethics forums where non-technical voices were not just invited, but essential.

It didn't feel like bureaucracy. It felt like shared ownership. A sign that the organization was growing up.

As the system scaled, Arjun understood another hard truth. Good intentions fade unless systems are designed to remember values for us.

So ethical intelligence moved from aspiration into architecture. Inspired by principles that emphasized purpose, respect, transparency, and explainability, fairness became something the system itself monitored. Models began flagging drift. Ethical risks surfaced alongside financial KPIs. Long-term societal consequences were modelled, not assumed away.

AI was now watching AI. Not to control it—but to keep human judgment in the loop.

The organization learned something quietly profound: ethics is not a checkpoint you pass. It is an ongoing conversation with the future.

Just as the bias issue began to stabilize, a second challenge surfaced.

A product leader said what many were thinking. "We can't compete. Our competitors have ten times more data."

This time, Arjun smiled.

He had learned that scarcity, while uncomfortable, often sharpens creativity. Abundance, on the other hand, can dull it.

Another myth had to go—that only massive proprietary datasets create successful AI. He began encouraging cognitive flexibility instead. Quality over quantity. Learning speed over data volume. Experimentation over perfection.

Rather than waiting for the "right" dataset, teams moved into sandbox environments. They experimented with synthetic data, transfer learning, augmentation techniques, and scenarios generated by domain experts. Some ideas failed quickly. Others worked in unexpected ways.

What mattered wasn't success or failure—it was the loop. Try. **Learn. Adapt. Repeat.**

Slowly, the questions changed. Teams stopped asking whether they had enough data. They started asking whether they knew enough to act responsibly. Sensemaking replaced anxiety. Learning replaced defensiveness.

Data scarcity pushed the organization outward as well. Partnerships formed with universities, research groups, and even non-competing firms. This required trust across cultures, a willingness to release "not invented here" thinking, and sensitivity to regional norms around data and privacy.

Federated learning allowed intelligence to grow without forcing data to move. Inclusion became architectural, not symbolic.

The partnership between humans and machines deepened. AI-generated possibilities at scale. Humans judged realism, consequence, and ethics. Together, they unlocked growth without sacrificing responsibility.

Looking back, Arjun shared a simple insight he had learned from watching other organizations succeed. They didn't wait for perfect conditions. They tested quickly. They prioritized wisely. They embedded learning into culture. And they governed before scaling.

Late one evening, as the office lights dimmed, the realization settled fully.

These were never technical problems.

There were leadership capability gaps.

Bias could not be solved by governance alone. Data scarcity could not be solved by technology alone. Each challenge demanded something integrated—ethical clarity to define what should be, sensemaking to understand what is, emotional intelligence to face discomfort, learning agility to evolve, cultural intelligence to collaborate, systemic thinking to turn improvement into a habit, and human–AI partnership to scale wisely.

Leadership maturity, he now understood, mattered more than data volume ever could.

And that understanding changed everything.

What the Reader Should Learn from This Story

1. **AI failures are rarely algorithmic—they are leadership blind spots**
2. **Bias and data scarcity are not blockers; they are maturity tests**

3. **Spiritual, ethical, and cognitive agility compound over time**
4. **The best AI leaders don't chase certainty—they build adaptability**
5. **Responsible AI is not slower AI—it's sustainable AI**

The Closing Insight

Arjun didn't become a better AI leader because he mastered models.

He became a better leader because he mastered **himself, his teams, and the systems they lived in**.

That is the real promise of integrated intelligence.

And that is how invisible barriers become **sources of enduring advantage**.

Story: 27

Accuracy Isn't Enough: Ethical Decision-Making in an AI-Accelerated World

> **"The greatest enemy of knowledge is not ignorance; it is the illusion of knowledge."**
>
> — *Stephen Hawking*

The first warning came from a spreadsheet.

Everything looked perfect.

Forecast accuracy was up. Costs were down. Customer complaints had "statistically normalized."

The board applauded.

But Anika couldn't sleep.

She wasn't an engineer. She wasn't a data scientist. She ran regional operations—close enough to the ground to notice when reality felt... off.

The model said demand in the southern markets would collapse within six months. The recommendation was clean and logical: close two distribution hubs, reallocate inventory, and reduce staff.

The numbers were flawless.

Too flawless.

Anika didn't argue. She asked for the *reasoning trail.*

The system responded with confidence intervals, correlations, and optimization paths. Efficient. Elegant. Empty.

She flew south anyway.

What she found wasn't declining demand—it was **changing behavior**. Customers weren't buying less; they were buying differently. Smaller orders. Faster cycles. Local substitutions. Signals the model smoothed away as "noise."

Back home, Anika didn't call the system wrong. She changed the workflow.

From then on, the model didn't make decisions. It made **proposals**.

Anything above a certain confidence threshold flowed automatically. Anything ambiguous was routed to **people** who understood context—sales leads, logistics planners, regional partners.

When they intervened, the system didn't just accept corrections. It asked *why*.

Patterns emerged.

The AI learned the shape of local nuance. The humans learned what the system could see across thousands of micro-signals.

Something shifted.

The system stopped chasing perfection. The people stopped fearing it.

Three months later, the same model flagged a cost-saving opportunity Anika would have missed entirely—but paused itself, tagging the recommendation with a warning: *Potential social impact detected.*

It didn't know what to do with that. So, it waited.

That pause saved a supplier network—and a reputation.

The board never noticed the moment things changed. They only saw the results: steadier growth, fewer reversals, better morale, faster innovation.

What they didn't see was the real redesign.

Not of software. Of **authority**.

The system no longer replaces judgment. It *respected* it.

People weren't optimized out. They were positioned where judgment mattered most.

And the most important line in the new operating manual wasn't technical at all:

If the system feels confident but reality feels wrong—stop.

Because intelligence without context is just speed. And speed without wisdom eventually crashes.

Anika still reviews the dashboards every morning.

But she no longer asks, *"Is this accurate?"*

She asks, *"What is this not seeing—and who should?"*

That question changed everything.

5 practical daily questions + actions you can apply directly from the story

1. **Where does the data look "too perfect"?**

   Question: What result feels right on paper but wrong in reality?

   Daily Action: Pause before acting on high-confidence recommendations. Ask one human closest to the work: *"Does this match what you're seeing?"*

2. **What decision should be a proposal, not an order?**

   Question: Which automated or habitual decision deserves human judgment today?

   Daily Action: Turn at least one "automatic" decision into a review point where context can enter.

3. **Who holds the missing context?**

   Question: Whose perspective is not represented in this insight or decision?

   Daily Action: Invite one frontline, cross-functional, or regional voice into the conversation before finalizing.

4. **Where should learning replace correction?**

   Question: If something was overridden, do we understand *why*—or did we just fix the outcome?

   Daily Action: Document one "why" behind a correction and feed it back into the process, system, or habit.

5. **What needs judgment, not optimization?**

   Question: Where might efficiency conflict with values, trust, or long-term impact?

   Daily Action: Explicitly name the trade-off and choose consciously—even if it slows things down.

   One-Line Daily Reminder

   Don't ask *"Is this accurate?"* Ask **"What is this not seeing—and who should?"**

Story: 28

Why Data Mastery Fails Without AI-SENSE² – A Human-Centered Way to Build AI That Actually Works

> **"Information is not knowledge. Knowledge is not wisdom."**
>
> — *Clifford Stoll*

For years, data mastery lived in a narrow, technical corner of organizations.

If you could collect large volumes of data, clean the mess, train a model, and deploy it into production, you were labelled "advanced." Success was measured in pipelines completed and dashboards shipped.

The assumption was simple: better data and better algorithms would naturally lead to better decisions.

Then AI entered the real world of organizations.

And a sobering truth began to surface.

AI was not failing because data was missing. It was failing because meaning was unclear, trust was broken, learning lagged behind change, ethics were sidelined, and culture quietly resisted.

This is where **AI-SENSE²** emerges—not as another framework, but as a shift in how data mastery itself is understood.

AI-SENSE² Leadership Map: Navigating AI Transformation

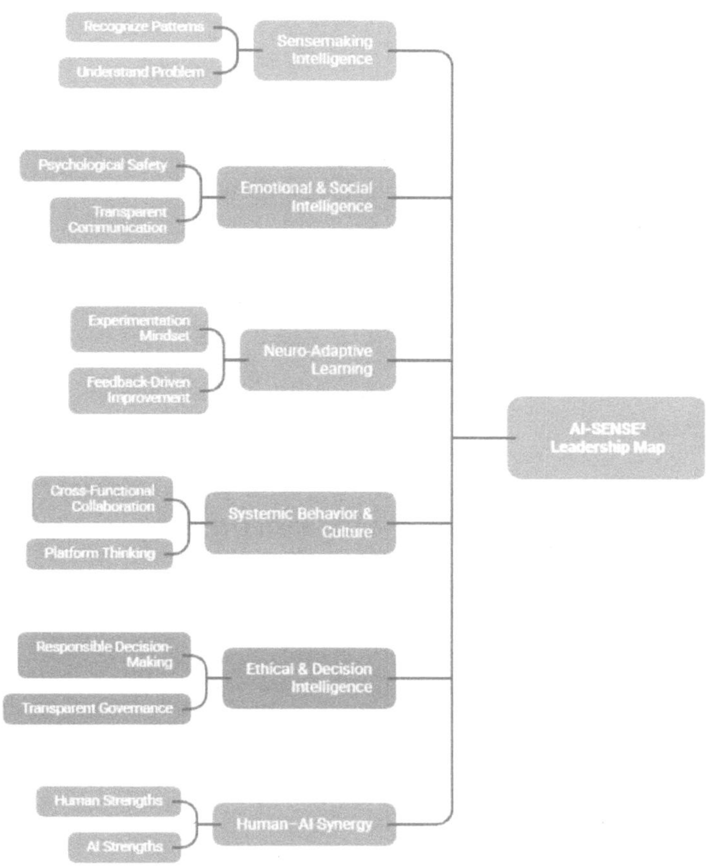

AI-SENSE² reframes data mastery as a leadership capability. One that blends human intelligence, neuroscience, ethics, and human–AI collaboration to create impact that actually lasts. It moves data out of spreadsheets and models and into the lived reality of how people think, feel, decide, and work together.

Sensemaking Intelligence: Finding Meaning in the Noise

Inside organizations, data is rarely messy only in a technical sense. It is socially fragmented.

Sales teams talk about churn one way. Support teams see it through a different lens. Product teams interpret the same numbers yet arrive at another story altogether.

A purely technical response tries to smooth these contradictions away—averaging, normalizing, and abstracting until the tension disappears. But the tension is the signal.

Leaders practicing Sensemaking Intelligence lean into the discomfort. They ask deeper questions: *Why does this data feel fragmented? What realities, incentives, and assumptions are shaping these numbers?*

They step away from dashboards and into dialogue. Definitions are surfaced, assumptions are challenged, and hidden incentives are exposed. Slowly, fragmented truths begin to align.

The breakthrough does not come from a better model. It comes from shared meaning.

Only when people agree on what the data truly represents does trust begin to form—and only then does AI earn the right to be believed.

Emotional & Social Intelligence: Turning Resistance into Trust

Data ownership is never neutral. It is emotional.

People rarely resist AI because it is wrong. They resist it because it threatens their identity, their expertise, and their hard-earned intuition. Beneath the surface of "data quality issues" often lie fear, loss of control, and unspoken anxiety.

AI-SENSE² leaders notice this early. They do not label resistance as defiance. They recognize it as fear, asking for safety.

Instead of imposing governance through policies and mandates, they begin with themselves—regulating their reactions, listening beneath the words, and reading emotional undercurrents. They create psychological safety and invite teams into co-creation rather than compliance.

Gradually, something shifts.

Governance stops feeling like control. Trust replaces defensiveness. And when trust returns, data starts flowing again.

Only then does AI become usable.

Neuro-Adaptive Learning: Keeping Pace with a Moving Landscape

The world of AI evolves faster than traditional learning models can keep up with. New tools, architectures, and platforms emerge constantly. Waiting to "fully learn" before acting is no longer an option.

Neuro-Adaptive leaders learn differently.

They do not memorize tools; they recognize patterns. They connect new technologies to existing mental models. They learn just enough, exactly when it is needed. And they transfer insights fluidly across domains.

They prototype early, learn while moving, and teach by doing. Knowledge is not stored—it flows.

In this environment, data mastery becomes a living capability. Not a static skillset, but an evolving intelligence that adapts as fast as the landscape itself.

Systemic Behavior & Culture: Turning Quality into Habit

Most organizations treat data quality as a cleanup exercise.

It appears once a quarter. After something breaks. Owned by a small group of specialists.

Systemic leadership asks a more uncomfortable question: *If data keeps degrading, what is the system teaching people to do?*

AI-SENSE² leaders understand that behavior beats policy. They design systems that make good data inevitable rather than optional.

Short data health checks become part of daily rituals. Shared ownership shows up in stand-ups. Reinforcement is embedded directly into workflows.

Over time, quality stops being a task. It becomes a habit.

And when habits change, systems begin to self-correct.

Ethical & Decision Intelligence: Choosing Wisdom Over Optimization

AI is brilliant at optimization. Leadership is about judgment.

Historical data carries historical bias. A model can be perfectly accurate and still deeply unfair. Ethical & Decision Intelligence gives leaders the courage to pause and choose consciously.

They recognize when efficiency collides with values. They make trade-offs under uncertainty with open eyes. They decide not just how AI *can* be used, but how it *should* be used.

Sometimes that means sacrificing a small gain in accuracy to protect trust, fairness, and legitimacy.

That is not a technical compromise. That is responsible leadership.

Intercultural × Human–AI Intelligence: Scaling with Humanity Intact

AI does not scale into a single culture.

What feels helpful in one region can feel intrusive in another. Privacy expectations differ. Communication styles vary. Trust in automation is shaped by history and context.

The squared dimension of AI-SENSE² expands data mastery beyond the individual system.

Leaders design collaboration instead of replacement. They balance global coherence with local adaptation. They build interfaces that respect cultural context.

AI contributes insight. Humans contribute meaning, judgment, and cultural translation.

That is how AI becomes resilient across borders, teams, and societies.

A New Way to Build AI

The traditional path to AI is linear:

Collect data. Clean it. Train a model. Deploy—and hope adoption follows.

AI-SENSE² takes a different path.

It creates shared meaning before modelling. Builds trust before enforcing governance. Learns faster than the technology changes. Engineers' habits instead of relying on rules. Makes ethical trade-offs deliberately. And designs for humans, cultures, and machines together.

The outcome is not just a functional AI system.

It is an AI system people believe in, adopt willingly, and sustain over time.

The Real Takeaway

Data mastery is no longer just a technical skill.

The most advanced AI solutions fail not because algorithms are weak, but because governance exists only on paper, trust is absent, culture resists transparency, ethics are treated as compliance, and learning cannot keep pace with change.

AI-SENSE² provides the meta-capabilities that make technical excellence matter in the real world.

Because the future of AI does not belong to the smartest machines.

It belongs to the wisest human systems— those that know how to think, decide, and lead alongside them.

To make these points resonate with leaders and stakeholders, we need to sharpen the prose to emphasize that AI is a **cultural mirror**, not just a software upgrade.

Here is a refined, "leadership-first" version of your five questions:

The Human-Centric AI Audit: 5 Questions for Wise Leadership

The success of an AI solution is rarely determined by the sophistication of the model. Instead, it is determined by the wisdom of the questions that shape it. Before deploying technology, leaders must audit the human landscape it will inhabit.

1. The Meaning Gap: Whose "Truth" is the Data Telling?

Before we talk about models, algorithms, or metrics, we must remember that every system is created and used by people. Humans bring their own experiences, biases, incentives, and interpretations to the data they work with. What one team calls "customer success," another team may define differently. Over time, these different meanings quietly get built into the same dataset.

When these definitions are not aligned, the problem becomes bigger once AI is introduced. Instead of creating clarity, AI can amplify the confusion that already exists in the data.

This is why leaders must pause and ask an important question: whose version of the "truth" is actually embedded in the data? What hidden assumptions are shaping the results that the system produces?

The real lesson is simple. If teams do not share a common understanding of the data and its meaning, even the most accurate models will not create real insight. Without shared meaning, accuracy alone has little value.

2. The Emotional Guardrail: Who Feels Exposed or Replaced?

Resistance to AI is rarely about the technology itself. In many cases, it comes from fear. When AI enters a workplace, it can challenge how people see themselves. It may question their expertise, their role in decision-making, or the value they believe they bring to the organization. If these emotions are ignored, people may not openly reject the system, but they may quietly resist it, slow it down, or avoid using it.

This is why leaders need to pause and ask deeper questions. What fears might this change be creating for the team? Are people worried about losing influence, control, or relevance? More importantly, are we creating an environment where people feel psychologically safe to learn, experiment, and contribute? Are we asking teams simply to follow the system, or are we inviting them to help shape it?

The key lesson is simple. Successful AI adoption is not just a technical challenge; it is a human one. When people trust the process and feel included in the journey, they are far more willing to share data, ideas, and insights. In the end, trust becomes the foundation that allows data—and innovation—to flow.

3. The Incentive Check: Are we Automating Bad Habits?

Data quality problems often look like technical issues, but in many cases, they are actually behavior problems in disguise. The way people enter, manage, and interpret data is shaped by the culture of the organization. If a company rewards speed over accuracy, or values "hitting the numbers" more than thoughtful reflection, those behaviors naturally show up in the data. When AI is introduced into such an environment, it does not fix those habits—it simply accelerates them.

This is why leaders need to step back and examine the behaviors their systems encourage. What actions are people truly rewarded for today? Are teams pushed to move quickly without verifying information? Are they encouraged to report results rather than question assumptions? These cultural signals directly influence the quality of the data that AI relies on.

The key insight is that AI will magnify whatever behaviors already exist in the organization. If the culture values careful thinking, transparency, and accountability, AI will strengthen those strengths. But if the culture rewards shortcuts or surface-level success, AI will expose and amplify those weaknesses.

4. The Value Compass: When is "Optimal" Actually Wrong?

AI systems are built to identify the fastest, cheapest, or most efficient way to reach a goal. They analyze data and optimize outcomes based on the rules and objectives we give them. But leadership is not only about efficiency. Leadership is about judgment—deciding whether the most efficient path is also the right one.

When organizations rely heavily on AI, ethical trade-offs do not disappear. They simply become less visible if leaders choose not to acknowledge them. A decision that looks statistically "optimal" may still conflict with fairness, responsibility, or long-term trust. Technology can recommend an action, but it cannot decide whether that action aligns with the organization's values.

This is why leaders must pause and ask deeper questions. Where might optimization clash with our core principles? Are there outcomes that we would refuse to accept, even if the data suggests they are efficient? These reflections are essential because algorithms can optimize processes, but they cannot define what is morally acceptable.

The essential lesson is clear. Efficiency is a powerful tool, but ethics is a human choice. True leadership lies in knowing the difference—and having the courage to act on it.

5. The Feedback Loop: Is the System Capable of Learning?

AI is not a "set-and-forget" product; it is a living system. As markets shift and contexts change, a static AI becomes a liability. If the feedback loop between human intuition and machine output is siloed, the system will fail.

- **The Audit:** How does feedback flow from the front lines back into the model? Who owns the **continuous learning** process, not just the initial launch?

- **The Bottom Line:** Sustainable AI is built on a partnership of continuous growth, not a one-time implementation.

Final Reflection

By asking these five questions, you shift the narrative. You stop treating AI as a technical hurdle and start treating it as a **leadership challenge.** The goal isn't just to build a more intelligent model; it's to build a more capable, transparent, and resilient organization.

Story: 29

Making AI Board-Ready: Turning Intelligence Into Enterprise Value

> **"The essence of strategy is choosing what not to do."**
>
> — *Michael E. Porter*

Every boardroom conversation about AI eventually arrives at the same unspoken question:

"Will this actually create value—or just complexity?"

Boards are not impressed by models, platforms, or demos. They are persuaded by clarity—clarity of value, clarity of risk, clarity of execution, and increasingly, clarity of *how humans and AI will work together*.

In the AI era, the difference between an approved initiative and a stalled one is not technical sophistication. It is whether the business case speaks the **language of enterprise leadership**.

This is the story of how leaders do that well.

What Makes an AI Business Case Truly Board-Ready?

A board-ready AI business case answers **three questions with discipline**:

1. **Will this materially increase enterprise value?**
2. **Are the risks—financial, ethical, operational—understood and governed?**
3. **Does this strengthen our long-term competitive and cultural position?**

Everything else—algorithms, vendors, architectures—is secondary.

But there is a deeper shift beneath these questions.

The strongest AI business cases do not just propose technology. They redefine **how people think, decide, and behave at scale**.

This is where many AI initiatives fail—and where successful ones quietly excel.

The Hidden Success Factor: Human–AI Synergy

The most successful organizations are not "using AI." They are **restructuring how intelligence flows through the enterprise**.

AI is no longer a tool at the edge. It is becoming a thinking partner—one that reshapes leadership, teamwork, and culture.

Boards increasingly look for this clarity:

- Who decides—and who advises?
- Where does AI augment judgment, and where does human accountability remain absolute?
- How do behaviors change so adoption actually sticks?

When leaders can answer these questions, trust follows.

From Insight to Approval: The AI Business Case Archetypes Boards Approve

Across industries, successful approvals consistently fall into eight archetypes. Each one aligns **economic value**, **human judgment**, and **systemic behavior change**.

When AI enters the boardroom, it rarely enters as a technology discussion. It enters as a value conversation. Directors are not asking whether a model works. They are asking whether it changes the trajectory of the business without destabilizing the enterprise that carries it.

The strongest AI growth stories begin with an unexpected shift in framing. AI is not positioned as automation, but as precision. Leaders explain how it sharpens focus—helping teams identify the right customers at the right moment with the right offer. Sales cycles shorten not because people work harder, but because effort is better directed. Conversion rates rise, deal sizes expand, and revenue scales without a corresponding rise in cost. What boards listen for is not excitement, but discipline. If this works, can it scale without eroding trust, quality, or brand?

Cost transformation follows a similar logic. The most credible cases do not talk about cutting headcount or extracting efficiency at any price. Instead, they describe operating leverage created by redistributing effort. AI absorbs routine, repetitive work, while humans focus on complexity, judgment, and exception handling. Quality improves alongside efficiency because people are no longer stretched thin doing work that dulls attention. Boards pay close attention to how leaders manage the transition—how incentives evolve, how workflows are redesigned, and how teams are supported through change. The real question is structural: does this strengthen margins while preserving resilience when things go wrong?

Then come the growth opportunities that were always visible but never reachable. Many markets were never unprofitable—only unscalable. AI changes the cost-to-serve equation, making it viable to reach mid-market segments, long-tail customers, or new geographies that once required disproportionate effort. Boards approve when leaders articulate not just near-term revenue, but strategic optionality.

Do these open doors that competitors cannot easily follow?

Does it create a foothold that compounds over time?

Not all value, however, arrives as growth. Some of the most persuasive AI cases focus on risk—specifically, the risks that never make headlines because they are prevented early. Predictive maintenance avoids outages before they cascade. Fraud detection stops losses before they become scandals. Compliance monitoring catches drift before regulators intervene. Boards respond strongly when leaders quantify losses avoided and clearly explain governance. The question they ask is stark: what happens if we don't do this?

In mature markets, attention often turns inward—to customers already earned. AI enables earlier detection of churn signals, more personalized intervention, and intelligent expansion of existing relationships. These cases resonate because the economics are fast, measurable, and defensible. Boards understand retention intuitively. If you can see customers leaving before they go, why would you let them?

In regulated industries, the conversation deepens. AI is no longer optional—it is existential. Compliance workloads grow faster than human capacity. AI reduces false positives, lowers manual review costs, and dramatically reduces regulatory exposure. But approval comes only when leaders demonstrate ethical safeguards, audit readiness, and the ability to scale responsibly as regulations evolve. Trust, not efficiency, becomes the asset being protected.

Time itself becomes the next strategic currency. In industries where speed determines winners, AI compresses discovery cycles, raises success rates, and enables earlier market entry. Boards respond when leaders connect AI directly to time economics. How much is time worth in this industry? And what is the cost of being late?

Sometimes, the strongest case is defensive. When competitors deploy AI, inaction becomes a measurable risk. Market share erodes quietly. Customers drift. Talent leaves for more advanced environments. Boards listen carefully when leaders quantify the cost of delay and name it honestly. What is the price of doing nothing?

Yet even the most compelling business case fails if adoption stalls. Board-ready leaders address this directly. They explain how AI reshapes behavior at every level of the organization. At the individual level, leaders model new habits—using AI to challenge assumptions rather than confirm bias, delegating cognition without surrendering accountability, and making ethical judgment explicit rather than implicit. At the team level, new norms emerge: decisions informed by AI but debated openly, rituals that reinforce learning, and psychological safety to question machine outputs. At the system level, culture is engineered deliberately—through incentives aligned to AI-enabled outcomes, reinforcement loops that sustain change, and adoption curves tracked with the same rigor as financial KPIs.

This is how transformation sticks.

At its best, AI does not replace leadership. It raises its standard. Machines handle analysis, prediction, and pattern recognition. Humans retain context, ethics, and accountability. Decisions become faster, clearer, and more resilient. Organizations that master this partnership do not merely deploy AI. They think better as a system.

Behind every approved AI initiative sits a familiar structure. A one-page executive summary that makes ROI and payback unmistakable. Clear strategic alignment. Financial models that include downside scenarios. Phased implementation with explicit stop-go points. Named accountability. Transparent risk, ethics, and governance plans. Metrics the board can track quarter by quarter. This structure does one essential thing: it reduces uncertainty. And uncertainty—not technology—is what boards fear most.

Before entering the boardroom, seasoned leaders ask themselves a final set of questions.

Can I defend every assumption?

Do I know the cost of being wrong?

Can we stop early if this fails and still learn?

Who is personally accountable?

Will this still matter in three years?

If those answers come calmly and clearly, the case is ready.

Because AI does not win board approval by being impressive. It wins by making people more capable, decisions wiser, and organizations stronger. Boards do not fund technology. They fund clarity, confidence, and value creation.

When AI strengthens human judgment, embeds new behaviors, and scales enterprise intelligence, approval is no longer a question of if.

Only how fast.

Omega Healthcare Management Services—an organisation supporting over 350 healthcare providers—faced a massive operational challenge. Their teams were manually processing **millions of insurance billing and medical documentation tasks**, consuming enormous time and energy.

They decided to integrate **AI-powered automation with human oversight** by partnering with UiPath's Document Understanding tools. Rather than replacing people, they redesigned workflows, so AI handled repetitive extraction and preparation while humans focused on exceptions and value-added work.

The results were significant:

- Over **100 million documents processed** through AI automation since 2020
- More than **15,000 employee hours saved every month**
- Documentation turnaround times **reduced by 40–50%**
- Accuracy reached **99.5%**, reducing error-related rework
- Clients experienced a **30% ROI from the automation program**
- Human roles shifted to more knowledge-driven tasks rather than rote processing

This is a clear example of **Human–AI synergy** in action: AI amplifies capacity and accuracy, while humans retain governance, decision quality, and contextual judgment. It's not AI replacing professionals — it's AI *augmenting* them.

Why This Story Matters for Board-Ready AI Business Cases

This real case illustrates several key themes that boards react positively to:

1. **Clear Value Creation (Enterprise Value)** Omega's AI initiative delivered measurable ROI and time savings, not just internal efficiency talk. Boards care about **tangible business impact**, not promises.
2. **Risk and Governance** Rather than pushing "full automation," leaders framed AI as a partner — well-governed, transparent, and supervised — reducing risk and maintaining quality.
3. **Human–AI Synergy:** AI didn't replace human roles outright. It *raised the quality and satisfaction* of human work by

removing mundane tasks and enabling employees to focus on more strategic and complex problems.

4. **Culture and Behavior Change** Rather than treating AI as a bolt-on tool, Omega redesigned workflows. Roles evolved. Teams adapted to new rhythms. This speaks to **systemic behavior and cultural adoption**, not isolated pilots.

"Omega Healthcare deployed AI not as a stand-alone tool but as a collaborator with human experts. Within months, it halved document turnaround times, saved thousands of hours of manual work, and delivered a clear ROI — while uplifting human roles toward higher-value work."

Part VII

Human–AI Synergy

Working with AI Without Losing What Makes Us Human

Introduction

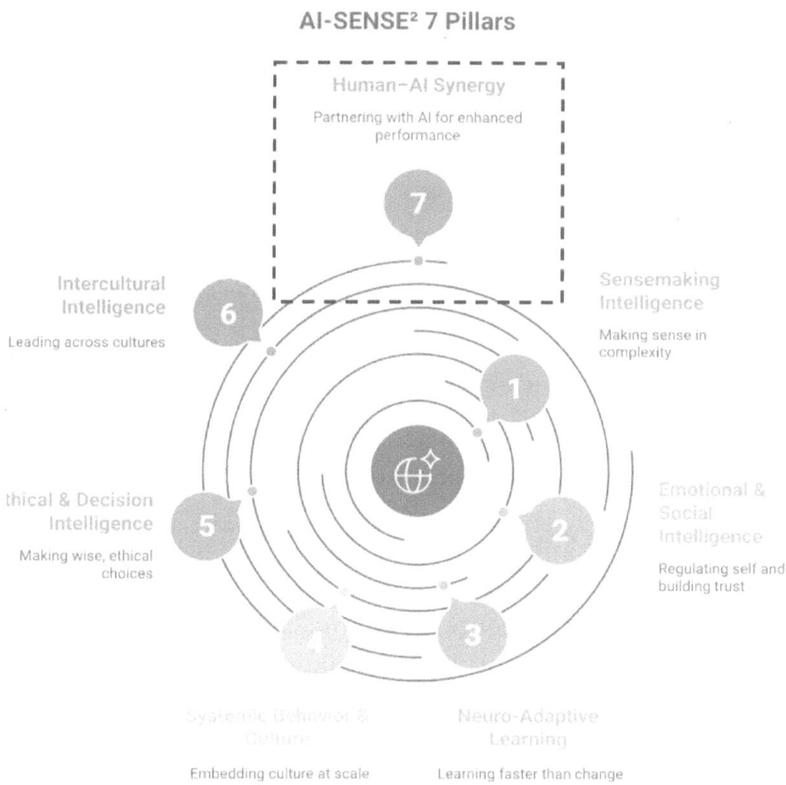

As artificial intelligence grows more powerful, excitement often dominates the conversation. Yet beneath that excitement lies a quieter and more profound question:

What happens to human judgment, intuition, and meaning when machines begin doing more of the thinking?

Part VII—**Human–AI Synergy** confronts this question directly. Rather than viewing AI through the familiar lens of competition or replacement, this section explores a more important shift: the movement from **human versus machine** to **human with machine**.

But this is not a romantic partnership with technology. It is a disciplined and intentional collaboration—one in which humans remain responsible for the decisions that shape organizations, societies, and lives.

For leaders, the challenge is no longer simply learning **how to use AI tools**. The deeper challenge is learning **how to work alongside intelligent systems without surrendering responsibility, wisdom, or humanity**.

Across these chapters, readers explore how AI can subtly weaken leadership when misused—encouraging cognitive laziness, blind delegation, or emotional distance from decisions. At the same time, they discover how AI can strengthen leadership when it is treated not as an authority, but as a powerful thinking partner.

This section examines critical questions such as:

- Why leadership presence becomes **more visible—not less—inside algorithmic systems**
- What forms of work remain **fundamentally human even in highly automated environments**
- How treating AI like a "brilliant intern" can quietly erode judgment instead of strengthening it
- Why intuition, experience, and moral awareness remain essential for stabilizing AI-driven decisions under uncertainty

At its core, this part reframes **intuition as a form of intelligence**—not something that competes with data, but something that helps interpret it wisely. When human wisdom meets machine capability, AI stops feeling abstract or threatening and begins to serve as a grounded, trustworthy partner.

By the end of Part VII, the question of whether AI will replace human leadership fades away. A more important realization emerges:

The quality of human leadership ultimately determines the quality of AI outcomes.

Outcome

Leaders learn how to build genuine **human–AI synergy**—working with intelligent systems in ways that amplify judgment, preserve meaning, and protect what makes leadership profoundly human.

Story: 30

The Human at the Heart of the Machine: Leading Across Cultures and Algorithms

> **"AI is the defining technology of our times, but it's human ingenuity and judgment that will determine its impact."**
>
> **— Satya Nadella**

When we imagined the future, we pictured quiet machines humming with perfect precision, AI predicting every risk, organizing every backlog, and leading teams with mathematical clarity.

But reality paints a richer picture—one where the **heartbeat of human leadership** has become louder, not softer, because of AI.

The Unforeseen Bottleneck

As AI tools swept through organizations, generating plans instantly, drafting user stories automatically, and summarizing meetings in seconds, relief initially set in.

We thought coordination would become effortless.

But a new, subtle bottleneck appeared. It wasn't technical or procedural—it was **human**.

- Teams still needed **direction** when priorities clashed.
- They needed **courage** when setbacks hit.
- They needed **trust** when the road ahead looked uncertain.

AI could calculate risk, but it could not calm a worried product owner.

AI could generate ten options, but it could not help humans make decisions they felt truly responsible for.

AI could produce reports, but it could not **inspire people to care.**

This was the moment your true human strengths—your **judgment, empathy, and courage**—shifted from being relevant to being **essential.**

The Leadership That AI Cannot Learn

Your role subtly transformed. You were no longer just the process facilitator; the tools could handle that. Instead, you became the anchor that stood in the middle of complexity and created **meaning**.

- When teams felt overwhelmed by the speed of automation, **you grounded them.**
- When AI generated ten viable delivery paths, **you helped them choose the one aligned with business reality.**
- When dependencies grew tangled despite powerful algorithms, **you brought the voice of reason and connection** back to the table.
- When people doubted themselves, **you helped them see possibilities beyond the cold metrics.**

Your work evolved from managing the process to **orchestrating intelligence**—human and machine, together.

The Art Within the Science

During a high-pressure product release, everything nearly collapsed. AI dashboards were flashing red, and models were predicting failure. Yet, it wasn't a tool that turned the situation around.

It was **you.**

You gathered the teams. You created a safe space for **honest conversation**. You simplified what seemed chaotic. You rebuilt the **trust** where fear had quietly settled.

In that moment, everyone realized something profound: **AI can guide action, but only a human can transform a situation.** Your calmness, your clarity, and your ability to connect people became the essential **anchor** that the organization didn't know it needed.

Your Unique Contribution

Your role has not shrunk; it has expanded. You are the:

- **Bridge** between the logic of AI and the emotion of humans.
- **Translator** of complexity and the **Designer of flow**.
- **Steward** of psychological safety and the **Guardian of value.**

The future belongs to the human who can lead the machine.

In an AI-powered world, the rarest and most valuable capabilities are precisely what AI cannot replicate: **judgment, empathy, connection, and courage.**

AI can accelerate work, but **you accelerate belief.** AI can generate plans, but **you produce teams who trust each other enough to execute them.**

This is your unique contribution—the part no model can ever replicate.

Here are the **top 3 key messages**:

- **AI amplifies human leadership—it does not replace it**

  The story reveals a quiet truth: as AI takes over coordination, planning, and analysis, **human leadership becomes more critical, not less**.

  AI can calculate risk, generate options, and surface insights—but it cannot provide direction when priorities clash, courage when uncertainty rises, or trust when teams hesitate.

Core message:

The more capable AI becomes, the more essential human judgment, empathy, and courage are.

- **The real bottleneck in the AI era is human meaning, not machine capability**

  Despite powerful AI tools, teams still struggle—not because the technology fails, but because **humans need clarity, reassurance, and shared purpose**.

  AI produces speed and options; leaders must create meaning, responsibility, and alignment.

  Core message:

  AI creates information. Leaders create meaning.

- **The leader's role has evolved from managing processes to orchestrating intelligence**

  The story shows a clear role shift:

  - From process facilitator → sense maker
  - From task coordinator → trust builder
  - From execution manager → orchestrator of human + machine intelligence

  In moments of pressure, it wasn't AI that saved the release—it was **human presence, calmness, and connection**.

  Core message:

  The leaders who thrive in the AI age are those who can blend machine insight with human wisdom to guide teams through complexity.

  In one line

  AI accelerates execution—but only human leadership can turn speed into trust, choices into commitment, and complexity into progress.

Story: 31

When Wisdom Met the Machine: How Human Intuition Turned AI from Ghost to Partner

> **"The real problem is not whether machines think, but whether men do."**
>
> — **B. F. Skinner**, psychologist and behavioral scientist

At the edge of the furnace floor stood Arjun.

Thirty years of heat had etched themselves into his posture. He didn't need dashboards or sensors. He could read molten steel the way a sailor reads the sea—by color, by sound, by instinct honed through decades of near-misses and quiet victories.

People said *smelting ran in his blood*.

So, when the company installed a multi-million-dollar AI system—screens glowing, algorithms humming—Arjun didn't see innovation.

He saw a ghost.

The data scientists from the city called it a *perfect model*. It promised optimized temperatures, reduced waste, and higher yields. But when the AI suggested a furnace setting that contradicted Arjun's gut, he ignored it. When it failed once—just once—he dismissed it entirely.

The machine didn't understand the steel. And worse—it didn't understand him.

Soon, the project bled money. Operators lost trust. Executives grew impatient. The "intelligent system" became little more than a flickering screen mounted high on the wall—present, expensive, and irrelevant.

The math wasn't wrong.

But the system had no soul.

A Different Kind of Intervention

Instead of pushing harder, leadership paused.

They stopped treating the plant like a collection of gears and began treating it like a **living organism**. That's when they introduced **AI-SENSE^2**—not as technology, but as a way of thinking.

The question changed from *"Why won't they use the AI?"* to *"What does the AI need to understand about being human?"*

Phase One: Breaking the Black Box

(Sensemaking & Emotional Intelligence)

The first shift wasn't technical—it was conversational.

Data scientists didn't arrive with laptops open. They arrived with notebooks closed. They sat beside Arjun during shifts and asked him how he decided when to adjust the heat. They listened as he spoke about sparks, vibrations, and that subtle moment when steel *"feels uneasy."*

Slowly, his intuition was translated—not replaced—into data.

At the same time, something deeper was addressed: fear.

No one had said it out loud, but Arjun felt it every day. *What happens to a man whose wisdom becomes an algorithm?*

Through deliberate emotional intelligence, leadership reframed the story.

"You're not being replaced," they told him. "You're being amplified."

The AI wasn't a judge. It was a lens.

And for the first time, Arjun leaned closer to the screen instead of turning away.

Phase Two: The Lab on the Floor

(Neuro-Adaptive Learning)

They abandoned one-time classroom training.

Instead, they created a sandbox—right on the shop floor.

Arjun and his team were encouraged to *break* the model. To push extreme scenarios. To see where it failed and why. The AI wasn't presented as a finished truth but as a learner—just like them.

Old habits surfaced. Some were useful. Some weren't.

And slowly, through deliberate practice, unlearning began.

The AI learned faster. But so did the humans.

Phase Three: The New Ritual

(Systemic, Ethical & Human–AI Intelligence)

The biggest change arrived quietly.

Shift handovers have transformed.

Instead of hurried notes and assumptions, teams reviewed decisions made jointly by humans and AI. They discussed where the AI helped—and where human judgment overruled it.

Ethics were explicit. Biases were questioned. And one rule was never broken:

The final decision always belonged to a human.

The AI became a teammate—not a master.

The Turning Point: From Skeptic to Sage

Something shifted in Arjun.

He no longer saw the AI as an enemy. He saw it as a thinking ally—one that could detect patterns invisible to the human eye, but blind to context, history, and deception when sensors lied.

Arjun became the bridge.

He translated data science into shop-floor language. He explained intuition to engineers. Younger operators listened to him—not because he resisted change, but because he embodied it.

Wisdom met intelligence.

The Intelligent Enterprise Emerges

The results showed up everywhere—but not all at once.

Fuel waste dropped. Throughput rose. Adaptation that once took weeks now happens in hours.

But the real victory wasn't in the numbers.

It was in trust.

Operators understood *why* decisions were made. The system learned continuously. And what began in one plant became a blueprint across many.

The organization didn't just use AI anymore.

It thought with it.

The Moral of the Story

Organizations don't fail at AI because of bad algorithms.

They fail because they try to install a digital brain into a system that has forgotten how to learn, feel, and adapt.

AI-SENSE^2 does something simple but powerful.

It doesn't "install" AI into an organization.

It changes how the organization thinks.

When that happens, humans and machines don't compete.

They learn from each other.

The machine stops being a mysterious ghost on the wall.

It becomes a working partner on the floor.

But that only happens if you ask the right questions.

Not fashionable questions.

Real ones.

1. What problem are we actually solving?

Most AI projects begin backwards.

A group gathers in a room and asks,

"Where can we use AI?"

That sounds ambitious.

It's usually vague.

The better question is:

“What is broken right now?”

Where are we losing time?

Where is quality unstable?

Where are customers frustrated?

Where are costs rising?

Where is risk increasing?

If you cannot describe the problem in one clear sentence, you don’t understand it yet.

And if you don’t understand it, adding AI will only add complexity.

Define the pain.

Define what improvement looks like.

Measure it in plain terms.

If an AI idea cannot clearly improve something real, drop it.

Enthusiasm is not a strategy.

Clarity is.

2. Who has to trust this system?

You can build an incredibly accurate model.

It won’t matter if no one uses it.

Adoption isn’t logical at first.

It’s emotional.

If frontline teams feel ignored, they will ignore the system.

If AI feels imposed, it feels threatening.

So, involve people before building the model.

Ask them how they actually make decisions.

What shortcuts do they use?

What patterns do they see?

Where they hesitate.

Translate that into the system.

Then explain the AI in language they already understand.

No jargon.

No mysticism.

When people see their own thinking reflected in the tool, trust begins.

Accuracy builds performance.

Participation builds trust.

And trust determines usage.

3. What judgment must remain human?

AI can process massive amounts of data.

It cannot carry responsibility.

That stays with people.

Automation without judgment is dangerous.

Fast mistakes scale quickly.

So, define clearly:

Where must a human review?

Where can someone override?

Who owns the decision?

If something goes wrong, "the system decided" is not an answer.

Someone must be accountable.

Train people not just how to use AI —

But when to question it.

A strong system combines:

Machine speed

Human judgment

Remove either one and risk increases.

4. How will this system adapt when reality changes?

Models decay.

Markets change.

Customers shift.

Regulations move.

Patterns evolve.

A static model slowly becomes wrong.

The goal isn't to build a perfect system.

It's to build one that learns.

Create feedback loops.

Compare predictions with outcomes.

Retrain models.

Adjust workflows.

Learning cannot be a one-time event.

It must be continuous.

Relevance is not achieved once.

It is maintained.

5. Are we building intelligence — or just buying software?

Purchasing tools is easy.

Building capability is harder.

If all knowledge sits with vendors or consultants, dependency grows.

Real intelligence lives inside the organization.

That requires roles:

People who translate business problems into technical ones.

People who operate and monitor systems.

People who maintain data quality.

People who make decisions.

Don't measure success by certificates.

Measure it by depth:

Can teams diagnose issues?

Challenge outputs?

Improve the system?

If yes, intelligence is growing internally.

If not, you are renting it.

6. Does AI live inside daily work?

Pilots are exciting.

Scaling is difficult.

If AI exists only in demos and presentations, it dies quietly.

It must appear in real work:

In handoffs.

In reviews.

In planning meetings.

In performance discussions.

Redesign workflows so humans and AI collaborate naturally.

Make usage visible.

Talk about failures openly.

Discuss improvements.

When AI becomes part of routine, it becomes normal.

And normal is powerful.

7. What behaviors must change?

Technology does not automatically change culture.

Old habits survive.

Fear blocks experimentation.

Blame blocks learning.

Control blocks exploration.

If mistakes are punished, people hide them.

If curiosity is discouraged, innovation slows.

Notice which behaviors prevent learning.

Then reinforce new ones:

Curiosity.

Transparency.

Thoughtful experimentation.

Reward insight — not just outcomes.

When behavior shifts, culture shifts.

Technology alone cannot do that.

8. What ethical risks are we scaling?

AI moves fast.

If something is biased or flawed, it spreads quickly.

Ethics cannot be postponed.

Ask early:

Is this fair?

Who could be harmed?

Who is accountable?

What happens if we are wrong?

Define guardrails before scaling.

Make decision rights clear.

Short-term gains mean little if long-term trust collapses.

Responsible systems scale safely.

Irresponsible ones cause damage.

9. Are leaders willing to change?

AI exposes leadership habits quickly.

If leaders demand certainty, experimentation stops.

If they react to early failures with panic, learning slows.

AI improves through cycles — not perfection.

Leaders must shift:

From control to enablement.

From answers to better questions.

From certainty to curiosity.

When leaders model learning, teams follow.

When leaders admit uncertainty, exploration increases.

In that environment, AI becomes a growth engine — not a threat.

The Unifying Question

Are we installing AI?

Or are we becoming an intelligent organization?

Because AI does not fail due to code alone.

It fails when:

Human wisdom is ignored.

Trust is absent.

Behavior remains unchanged.

AI works when:

Human judgment is designed into it.

Participation builds trust.

Learning becomes continuous.

Technology is only part of the system.

People are the rest.

If you leave them out of the equation, the model is incomplete.

And incomplete systems eventually fail.

The One Question That Ties Everything Together

Are we building an AI system—or becoming an intelligent organization?

Because successful AI deployment is not about installing software. It's about **how humans think, learn, decide, and adapt with machines**.

When these questions are asked—and acted on—*before deployment*, AI stops being risky.

It becomes transformative.

Here are the **top 3 key takeaways**:

- **AI fails when it ignores human wisdom—and succeeds when it learns from it**

  The AI didn't fail because the model was wrong.

  It failed because it dismissed decades of lived experience.

  Real impact began only when Arjun's intuition, judgment, and tacit knowledge were **translated into the system instead of overwritten by it**. AI becomes powerful not by replacing human expertise, but by **absorbing and amplifying it**.

  Takeaway: AI must learn from humans before humans can trust AI.

- **Trust is built through participation, not accuracy**

  The model was accurate long before anyone used it.

  Trust appeared only when people were invited into the process—listened to, respected, and protected from fear.

  By addressing emotional resistance, explaining decisions in human language, and keeping humans accountable, the AI shifted from a threat to a teammate.

  Takeaway: Adoption is emotional first, technical second.

- **AI transformation is a behavior change, not a technology upgrade**

  Nothing truly changed until daily rituals, decision ownership, and learning habits changed on the shop floor.

  When AI was embedded into real work—handoffs, reviews, and conversations—it stopped being a "project" and became **how work was done**.

  Takeaway: AI succeeds when organizations change how they behave, not just what they install.

  In one sentence:

  AI delivers value only when human judgment, trust, and learning are designed into the system—not bolted on afterward.

Story: 32

Coaching AI-Agile Leaders: Developing Human Intelligence in an AI-Driven World

> **"Wisdom is knowing what to do next; skill is knowing how to do it; virtue is doing it."**
>
> — **David Starr Jordan**

Coaching someone for an AI-powered workplace starts by helping them understand their current situation.

Everyone uses AI differently.

Some depend on it every day, some try it carefully, and some still avoid it.

Coaching begins by looking at what tools they use, what tasks they give to AI, and how confident they feel working with intelligent agents.

Once this picture becomes clear, it is easy to see which tasks AI can help with and which ones still need human judgment, creativity, and ethics.

After this foundation is set, coaching moves into developing the skills needed for AI-Agility.

The first skill is understanding how to use **AI responsibly.**

The coach helps the person learn what AI can do well, where it may fail, and when human oversight is essential. They learn how to set rules for AI and how to define "good output" before assigning a task. The aim is a safe, smart partnership — not blind trust.

As the person becomes better aligned with AI, they start experiencing how AI can truly amplify their abilities.

Coaching here helps them see AI not just as a tool that saves time, but as a partner that can broaden their thinking, speed up their work, and open creative possibilities.

They start identifying which parts of their job can be done faster or better with AI as support.

Because AI changes so q**uickly, learning agility becomes essential.**

The coach encourages curiosity instead of fear. New features, mistakes, or strange outputs become learning opportunities. Every interaction with AI becomes a moment to adjust, try again, and grow.

These fast changes can create emotional reactions.

That's why emotional regulation is at the center of AI-Agility.

The coach helps the person notice how they feel when AI does something brilliant or makes a big error. Instead of reacting with frustration, insecurity, or stress, they learn to pause, understand the feeling, and respond calmly and wisely. This builds emotional maturity.

Awareness and mindfulness add another layer of stability.

In a workflow filled with agents, small decisions matter — when to stop the agent, when to let it continue, and when something feels wrong.

Coaching strengthens the person's ability to stay present, pay attention, and sense the quality of what AI is producing. This awareness becomes their internal guide.

With these basics in place, coaching moves into cognitive leadership. Here, the person learns to switch thinking modes smoothly — being analytical when reviewing AI output, creative when pushing AI for better ideas, and ethical when facing questionable suggestions. A good coach helps the person grow this mental flexibility. While AI can provide answers, only humans can blend logic, values, and context.

Social intelligence becomes more important as AI takes over routine tasks.

Human work becomes more about relationships.

The coach encourages the person to notice emotional cues, understand team dynamics, and build deeper connections. With AI saving time, they can invest more in people, using empathy and communication to create a positive impact.

Ethics becomes the compass that keeps everything safe and grounded.

Since AI speeds up both good and bad actions, the coach helps the person develop strong moral clarity.

They learn what they will and will not allow AI to do for them. They begin to see themselves not just as users of AI, but as protectors of ethical behavior.

As all these abilities come together, coaching shifts toward helping the person perform at their best.

They learn how to enter a new kind of flow — where their own thinking, AI's outputs, and team collaboration all blend smoothly. They are no longer just responding to AI; they are leading it, like a conductor guiding an orchestra.

However, great capability also brings new pressure. The person may feel tired from checking AI outputs, feel pushed to produce more because AI is available, or feel overwhelmed by many digital channels.

Coaching on resilience helps them protect their energy, avoid burnout, and manage their digital environment wisely.

Finally, coaching supports the person in developing a strong digital focus — the ability to think deeply in a world full of fast AI responses and constant distractions. They learn when to pause AI, think on their own, and solve complex problems with full attention.

When all these elements come together, the person becomes an AI-Agile leader.

AI is no longer a separate tool — it becomes part of how they think, work, and lead.

Their human strengths — judgment, creativity, emotional intelligence, and ethics — shape the value AI can produce. They step into a stronger, clearer professional identity: someone who not only works with AI but uses it to create outcomes that neither human nor AI could achieve alone.

Here are the **top 3 key takeaways**:

- **AI-Agile leadership starts with self-awareness, not tools**

  Effective coaching begins by helping leaders understand *how* they currently use AI, *where* they rely on it, and *what* emotions and beliefs they bring into the relationship. Clarity about tasks that belong to AI versus those that require human judgment, creativity, and ethics is the foundation of responsible and effective AI use.

  Core message:

  AI-Agility is built by knowing yourself in relation to AI, not by adopting more technology.

- **Human intelligence becomes more valuable as AI capability increases**

  As AI accelerates execution, the differentiators shift to emotional regulation, mindfulness, cognitive flexibility, social intelligence, and ethical clarity. Coaching develops these human capacities so leaders can interpret AI outputs wisely, make context-aware decisions, build relationships, and act as ethical stewards rather than passive users.

Core message:

The more powerful AI becomes, the more leadership depends on distinctly human strengths.

- **Coaching transforms AI from a productivity tool into a leadership amplifier**

When learning agility, emotional resilience, ethical judgment, and focus come together, leaders enter a new kind of flow—one where human thinking, AI insight, and team collaboration reinforce each other. AI is no longer something they respond to; it becomes something they *lead*.

Core message:

AI impact scales when leaders learn to conduct intelligence—human and artificial—into one coherent system.

Part VIII

Intelligence at Enterprise Scale

From pilots to platforms, from experiments to advantage

Introduction

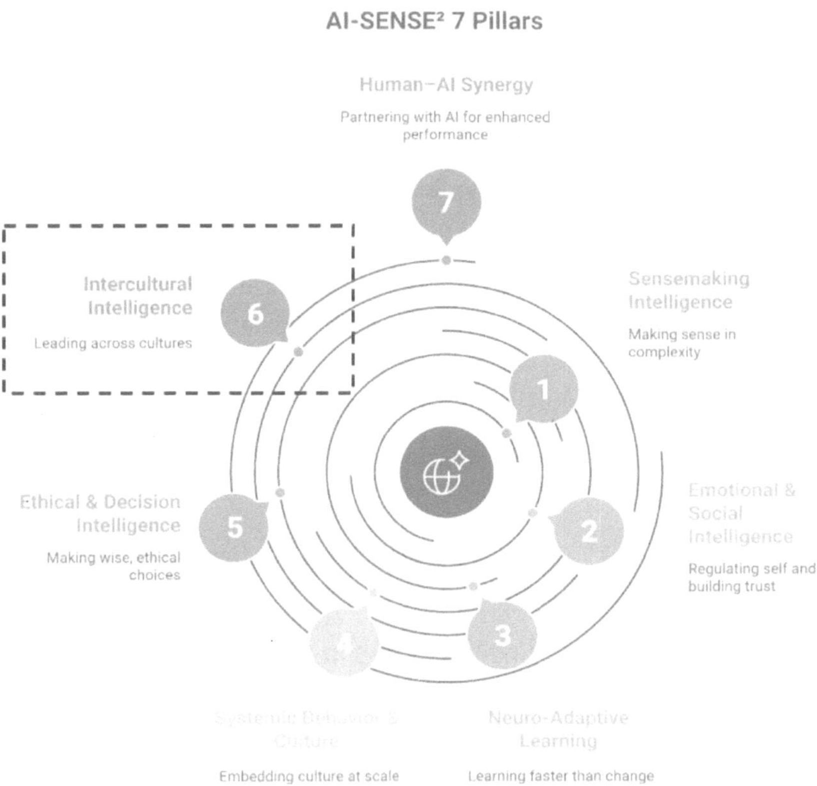

Many organizations have learned how to make AI work in small pockets of the business.

Very few have learned how to make it work across the entire enterprise.

Part VIII—**Intelligence at Enterprise Scale** explores one of the most difficult transitions in the AI journey: moving from promising pilots and isolated success stories to **enterprise-wide, repeatable intelligence**. This is the stage where early excitement often fades, complexity begins to multiply, and leadership capability is truly tested.

At scale, AI is no longer simply a technical initiative.

It becomes an **organizational intelligence challenge**.

Decisions now move across functions, geographies, and cultures. Accountability becomes more distributed. Speed increases while clarity often decreases. What once worked for a single team or use case begins to break down when multiplied across thousands of people, processes, and systems.

This is where many AI strategies quietly stall—not because the algorithms fail, but because the **human system surrounding them struggles to adapt**.

This section takes readers inside real large-scale transformations to reveal what actually allows intelligence to scale. Instead of focusing on hype or technical novelty, these chapters examine the leadership disciplines that enable global organizations to build sustainable human–AI collaboration.

Readers will explore:

- Why scaling AI requires **organizational alignment as much as technological capability**
- How decision-making changes when intelligence operates across complex global systems
- Why many organizations encounter a **"human wall"** when attempting to expand AI initiatives

- And how leadership architecture determines whether AI becomes a strategic advantage or an expensive experiment

Within this context, **AI-SENSE²** is introduced as the operating mechanism that connects strategy to execution—helping leaders translate ambition into consistent enterprise outcomes.

This part also brings AI into the **boardroom**, examining how leaders transform intelligent systems into enterprise value while managing risk, maintaining trust, and strengthening governance at the highest levels.

By the end of Part VIII, one insight becomes unmistakably clear:

Scaling AI is not primarily about deploying more technology.

It is about scaling judgment, trust, and alignment across the enterprise.

Outcome

Leaders learn how to move from **isolated pilots to enterprise platforms**, turning AI experimentation into sustained competitive advantage through the disciplined application of **AI-SENSE²**.

Story: 33

From Packages to Intelligence: How AI Is Transforming Global Logistics

AI in Global Logistics: Lessons from Large-Scale Package Networks

The following case illustration draws on publicly discussed developments in the global logistics industry. It reflects patterns observed across major parcel delivery companies that have adopted artificial intelligence, automation, and advanced analytics to improve large-scale operations. It is presented for educational purposes and does not represent confidential internal information from any specific organization.

The Growing Complexity of Global Logistics

Modern logistics networks operate at an extraordinary scale. Every day, millions of packages move through sorting facilities, transportation hubs, and delivery routes across continents.

For large global carriers, even small operational errors can ripple into delayed shipments, dissatisfied customers, and increased operational costs. As e-commerce expanded rapidly, logistics companies faced mounting pressure to move packages faster while maintaining extremely high levels of accuracy.

Traditional manual sorting processes—relying heavily on human scanning, routing decisions, and conveyor systems—began to strain under this scale. Seasonal surges, such as holiday shopping, created additional pressure on warehouses and distribution hubs.

Across the industry, companies began exploring how artificial intelligence and automation could support more reliable and scalable operations.

The Challenge: Managing Volume, Speed, and Accuracy

Large logistics networks must manage three competing priorities simultaneously:

- **Speed**, to meet customer expectations for rapid delivery
- **Accuracy**, to ensure packages are routed correctly
- **Reliability**, to maintain consistent service during peak demand

Manual handling processes can struggle when package volumes increase dramatically. Misrouted parcels, delayed processing, and warehouse bottlenecks become more likely during high-traffic periods.

To address these challenges, many logistics providers began investing in AI-enabled systems that combine computer vision, machine learning, and automated handling technologies.

AI-Enabled Sorting and Routing

In modern distribution hubs, advanced systems can scan packages rapidly using high-speed imaging and optical character recognition technologies. These systems read shipping labels, identify destination information, and route packages through automated conveyor networks.

Machine-learning models analyze historical shipping patterns to determine efficient routing decisions, helping facilities handle high package volumes more consistently.

Automation technologies assist with physical sorting tasks, directing packages toward appropriate loading areas while providing real-time tracking data for operational teams and customers.

These capabilities allow distribution centers to process packages at significantly higher speeds while maintaining greater accuracy than traditional manual systems alone.

Enhancing Visibility and Customer Experience

Artificial intelligence is also being used to improve package tracking and delivery predictions.

Predictive analytics models can estimate delivery times based on historical data, transportation conditions, and operational capacity. When delays occur, anomaly detection systems can flag shipments that may require intervention.

Customers benefit from improved tracking visibility and proactive notifications, while operations teams gain early insight into potential disruptions.

Implementing AI at Scale

Introducing AI into complex logistics environments typically requires a phased approach.

Many organizations begin with pilot projects in selected distribution facilities. These pilots allow teams to test the technology, refine operational workflows, and collect employee feedback before expanding the system to additional locations.

As systems mature, AI capabilities are integrated with warehouse management software, transportation planning tools, and customer tracking platforms.

Employee training also plays a critical role. Workers learn how to collaborate with AI-enabled systems, supervise automated processes, and manage exceptions that require human judgment.

Over time, continuous improvement cycles allow systems to learn from operational data and adapt to seasonal demand patterns, transportation disruptions, and evolving customer needs.

Organizational Impact

When implemented successfully, AI-supported logistics systems can generate multiple benefits:

- Faster package processing
- Improved routing accuracy
- Reduced operational bottlenecks
- Better delivery predictions
- Enhanced customer visibility

However, technology alone rarely drives these outcomes.

Successful transformation typically depends on leadership decisions about how technology integrates with human expertise.

Human-AI Collaboration in Logistics

Across the logistics industry, organizations have learned that AI works best when it complements human capability rather than replacing it.

Automation systems handle high-volume repetitive tasks such as scanning, sorting, and data processing. Human workers focus on exception handling, operational oversight, and complex decision-making.

New roles often emerge as systems evolve, including specialists who supervise automated workflows, investigate anomalies, and ensure quality control.

In this environment, AI becomes an operational partner rather than a substitute for human expertise.

Change Management and Workforce Development

Large-scale technology transformations also require careful change management.

Employees often have understandable concerns about how automation may affect their roles. Organizations that succeed with AI adoption typically invest in transparent communication, workforce training, and opportunities for employees to develop new skills.

Training programs may include role-specific learning modules, hands-on workshops, or peer mentoring initiatives designed to help employees adapt to AI-assisted workflows.

When workers understand how technology supports their work, adoption tends to increase significantly.

Continuous Learning Systems

Another important lesson from AI adoption in logistics is that systems must evolve continuously.

Machine-learning models improve as they process new operational data. Feedback from frontline employees helps identify areas where automation can be refined or adjusted.

In this way, the organization and the technology improve together.

AI systems are not treated as finished products. They become part of a living operational ecosystem that evolves over time.

Leadership Lessons from AI Adoption

The experience of logistics organizations adopting AI highlights several leadership principles:

- defining clear operational problems before introducing new technology
- investing in infrastructure and data quality
- supporting employees through training and communication
- maintaining human oversight in critical decisions
- fostering cross-functional collaboration across operations, IT, and analytics teams
- continuously refining systems based on real-world feedback

These practices help organizations move beyond isolated technology experiments toward sustainable enterprise transformation.

A Broader Insight

Artificial intelligence alone does not transform complex organizations.

Real transformation occurs when leadership, technology, and human capability evolve together.

In large logistics networks, AI helps process vast volumes of information and automate repetitive tasks. Humans provide judgment, oversight, and the ability to manage unexpected situations.

When these capabilities are combined effectively, organizations can achieve improvements in efficiency, reliability, and customer experience.

Note: This case example reflects publicly discussed trends in logistics automation and AI-enabled operations and is presented for educational and illustrative purposes.

Key Takeaways: Intelligence at Enterprise Scale

- **Pilots Prove Possibility — Platforms Create Advantage**

  Many organizations successfully run AI pilots, but competitive advantage comes only when AI capabilities are scaled into platforms that support operations across the entire enterprise.

  Small experiments show what is possible.

  Enterprise platforms make it repeatable.

- **AI Must Be Embedded in Core Operations**

  AI creates real value only when it becomes part of daily workflows—sorting packages, optimizing routes, predicting demand, or managing exceptions.

  When AI lives outside operations, it remains a lab experiment.

  When it integrates into processes, it becomes operational intelligence.

- **Data Infrastructure Determines Scalability**

  Enterprise-scale AI requires strong data foundations: integrated systems, standardized data, reliable pipelines, and real-time visibility.

Without clean and connected data, AI remains fragmented and difficult to scale.

- **Human–AI Collaboration Is the Engine of Scale**

  Automation alone cannot manage complex operations. Humans must remain involved in decision-making, exception handling, and system oversight.

  The organizations that scale AI successfully design workflows where:

  AI handles volume and pattern recognition.

  Humans apply judgment and contextual insight.

- **Continuous Learning Turns AI into a Living System**

  AI systems improve as they process new data and feedback from operations.

  Organizations that scale AI treat it as a **continuous learning capability**, not a one-time deployment.

  The platform learns.

  The workforce learns.

  The organization evolves.

- **Cross-Functional Alignment Is Essential**

  Scaling AI requires coordination across operations, technology, analytics, and leadership.

  Enterprise advantage emerges when these groups work together around shared outcomes rather than operating in silos.

- **Trust and Transparency Enable Adoption**

  Employees must trust AI systems before relying on them.

  Explainable insights, human oversight, and clear accountability ensure that AI supports decision-making rather than replacing it.

  Trust is the bridge between pilot success and enterprise adoption.

- **Leadership Turns Technology into Transformation**

  The final step from experimentation to advantage depends on leadership.

  Leaders must:

  - Define clear strategic goals
  - Invest in infrastructure and talent
  - Guide cultural change
  - Scale learning across the organization

  Technology enables transformation.

  Leadership makes it sustainable.

Story: 34

When Machines Begin to Whisper: Leadership Lessons from AI-Augmented Engineering.

From Signals to Insight: How AI Is Reshaping Aircraft Maintenance

The following narrative reflects widely discussed developments in the aerospace industry as companies integrate artificial intelligence into predictive maintenance systems. The example is presented as an illustrative scenario rather than a description of the internal operations of any specific organization.

When Prediction Alone Was Not Enough

For decades, aircraft maintenance depended heavily on human expertise.

Inside maintenance hangars and operational control rooms, engineers relied on experience, inspection, and judgment. Skilled technicians listened to engines, examined components, and interpreted subtle signals that indicated whether something might fail.

Experience was the compass.

Intuition often provided the edge.

As digital technologies matured, artificial intelligence began entering the field—not as a replacement for engineers, but as an additional analytical capability.

Sensors embedded throughout modern aircraft could collect enormous volumes of operational data: vibration patterns, temperature changes, pressure variations, and performance fluctuations across thousands of flights.

Machine-learning models could analyze these signals and identify patterns that suggested potential mechanical issues.

At first, many organizations assumed the primary value of AI would be prediction.

But the deeper transformation turned out to be something else entirely.

When Machines Began to Reveal Hidden Signals

Advanced analytics systems were able to detect extremely subtle patterns within aircraft data.

Some signals involved changes too small for humans to reliably identify, such as:

- slight shifts in vibration behavior
- very small temperature deviations
- micro-variations in pressure signals

When analyzed across thousands of aircraft and millions of flight hours, these signals began to reveal meaningful patterns.

AI systems could process this enormous scale of information continuously.

But crucially, the systems did not replace engineers.

Instead, they expanded human perception.

Machines handled scale, pattern recognition, and continuous monitoring.

Engineers focused on context, interpretation, and judgment.

The technology acted less like a replacement and more like a **perceptual amplifier**.

Building Trust Between Humans and AI

Early deployments were not perfect.

Some systems generated too many alerts. Engineers questioned their reliability. In certain cases, teams initially ignored the recommendations.

Rather than forcing adoption, organizations gradually improved the systems through feedback loops.

Maintenance teams reviewed AI-generated alerts and provided real-world outcomes. Engineers flagged false positives and identified edge cases that the models had not yet learned to interpret correctly.

Over time, the models improved.

Alert accuracy increased.

False alarms declined.

More importantly, trust began to grow.

This process reflected not only machine learning, but **organizational learning**—humans and algorithms improving together.

From Reactive Maintenance to Calm Foresight

As predictive systems matured, the nature of maintenance work began to change.

In the past, unexpected equipment failures often triggered emergency responses.

Technicians were called in urgently. Repairs were conducted under pressure. Operations teams worked to minimize disruption.

With better predictive signals, many of these events could be anticipated earlier.

Instead of reacting to failures, teams could plan maintenance interventions in advance.

The result was not simply improved efficiency.

It was a shift toward **more stable decision-making environments**, where engineers had time to analyze options carefully rather than responding to immediate crises.

Making Complex Decisions More Transparent

Another important development involved the transparency of AI recommendations.

Advanced systems increasingly provided contextual information alongside predictions, such as:

- probability of potential failure
- estimated time windows for intervention
- potential operational impact
- cost comparisons between planned and unplanned maintenance

This context helped engineers evaluate trade-offs more effectively.

AI provided analytical insight.

Humans remained responsible for final decisions.

Maintaining this balance proved important, particularly in industries where safety considerations are paramount.

A Cultural Shift Inside Engineering Teams

Beyond technical improvements, organizations also observed subtle cultural changes.

When predictive insights became more transparent and explainable, engineers felt more comfortable questioning both the system and their own assumptions.

Instead of viewing AI as a threat, many teams began treating it as a collaborative tool.

Investigating unusual signals became part of normal work rather than an exception.

Gradually, attention shifted from fixing visible failures to identifying early warning signals that might otherwise remain hidden.

In this way, AI expanded awareness rather than diminishing human expertise.

Designing Systems Around Safety and Oversight

In aviation and other safety-critical industries, human oversight remained central.

AI systems typically support decision-making rather than replacing it.

When risk levels were uncertain or potential safety implications were high, issues were escalated for human review.

This approach ensured that technological capability developed alongside strong governance and accountability.

Ethical design principles were not simply documented; they were embedded into operational processes.

From Data Overload to Focused Attention

One of the most valuable contributions of modern AI systems is the ability to filter overwhelming volumes of operational data.

Instead of presenting engineers with thousands of raw signals, intelligent systems prioritize issues based on relevance and potential impact.

This shift protects one of the scarcest resources in complex organizations:

attention.

By highlighting the most meaningful signals and reducing informational noise, AI systems allow experts to concentrate on solving the problems that matter most.

The Deeper Lesson

The transformation of aircraft maintenance illustrates an important insight about artificial intelligence.

The true value of AI rarely lies in prediction alone.

Its real power emerges when organizations learn how to combine machine intelligence with human expertise.

Machines can detect patterns at an extraordinary scale.

Humans provide judgment, responsibility, and ethical oversight.

When these capabilities work together, the result is not simply automation.

It is **augmented intelligence**.

Leadership Lessons from the Shift

Several leadership capabilities become visible in organizations navigating this transformation:

Adaptive Thinking

Leaders move from reactive maintenance models toward predictive awareness.

Continuous Learning

Systems improve through ongoing feedback between engineers and AI models.

Emotional Adaptability

Teams learn to manage uncertainty as technology introduces new insights.

Human-Centered Decision Design

AI supports decisions, while responsibility remains with human experts.

Focused Attention

Technology helps organizations concentrate on the most meaningful signals rather than overwhelming data.

Operational Resilience

Predictive systems reduce emergency interventions and support more stable operations.

Ethical Governance

Human oversight remains central, particularly in safety-critical industries.

The Question for Modern Leaders

Across many industries today, machines are already generating signals about performance, risk, and opportunity.

The challenge for leaders is not simply deploying artificial intelligence.

It is learning how to **interpret those signals wisely**.

In many organizations, the most important insights are already present in the data.

The real question becomes:

Are we listening carefully enough to hear them?

Key Learnings About AI Agility from the Story

Learning 1: AI Agility Is a Human Capability, not a Tool

The company didn't "install AI."

They **rewired how humans think, learn, and decide**.

AI made agility possible — humans made it meaningful.

Learning 2: Agility Comes from Weak Signals, Not Strong Events

Failures don't start as crises.

They start as whispers.

AI Agility is the ability to act early, calmly, and intelligently.

Learning 3: Speed Without Emotional Regulation Is Dangerous

Early resistance wasn't technical — it was emotional.

AI Agility requires leaders who can manage fear, ego, and uncertainty.

Learning 4: Learning Velocity Beats Expertise

The best engineers weren't the ones with the most experience.

They were the ones who learned fastest with AI.

In an AI world, adaptability beats seniority.

Learning 5: AI Agility Protects Human Attention

By filtering noise, AI restored focus.

The future advantage isn't more data — it's better attention.

Learning 6: Ethics Are the Accelerator, Not the Brake

Trust increased only when safety and transparency were non-negotiable.

Ethical clarity made adoption faster, not slower.

Story: 35

From Experimental Pilots to Enterprise-Scale Advantage: The Microsoft Blueprint

Microsoft's transformation into an AI-first enterprise is often reduced to headlines about its partnership with OpenAI or the launch of Copilot.

But the real story is not about models.

It is about leadership, culture, and operational discipline.

Microsoft's journey shows that scaling AI is not a technical challenge to solve — it is an organizational shift to manage. By moving from isolated AI experiments to a unified platform strategy, Microsoft turned AI from a speculative innovation into a repeatable competitive advantage.

I. The Foundation: Culture as the Operating System

Before AI could scale, the mindset had to scale.

Under CEO Satya Nadella, Microsoft reshaped its internal culture — not to prepare for AI specifically, but to prepare for change itself.

1. From "Know-It-All" to "Learn-It-All"

Traditional organizations reward expertise and certainty.

Microsoft intentionally replaced this with a growth mindset culture.

Curiosity was rewarded.

Learning was visible.

Admitting "I don't know" became acceptable.

Impact:

When AI tools were introduced, they were not perceived as threats to expertise. Instead, employees saw them as collaborators — tools that could amplify capability rather than replace identity.

Leadership signal:

Executives openly acknowledged early model limitations. This transparency reduced fear of the "black box" and encouraged experimentation at every level of the company.

The culture normalized learning before AI required it.

2. Purpose Before Technology

Microsoft avoided the common trap of "AI for AI's sake."

Each major AI initiative was anchored to a specific, painful business problem.

They did not ask, "What can AI do?"

They asked:

- How do we help developers overcome the blank-page problem?

  → GitHub Copilot

- How do we simplify complex data analysis for business users?

  → AI-enhanced Power BI

- How do we remove repetitive administrative burden?

  → Microsoft 365 Copilot

AI was deployed as a solution to friction — not as a demonstration of capability.

Purpose-created alignment.

II. The Architecture of Trust: Human-Centric Design

Enterprise AI does not scale without trust.

Microsoft understood that performance metrics alone would not drive adoption. Trust had to be embedded into the design.

3. The "Co-Pilot" Principle

The name was intentional.

A pilot is in control.

A co-pilot assists.

Across Microsoft products, AI suggests — humans decide.

Developers review Copilot's code.

Managers review AI-generated reports.

Users approve or edit content before finalizing.

This preserved accountability.

It also created feedback loops. Each correction improved the system.

AI was positioned as augmentation, not automation.

4. Responsible AI as an Accelerator

Some companies treat ethics like a box to tick.

Microsoft did not.

They built ethics into the system from the beginning.

They created Responsible AI and review boards.

They tested models for bias and fairness.

They made AI explainable, so users could see why a suggestion was made.

Customers could understand how decisions worked.

Enterprise clients could trust the governance system.

Because ethics was built early, risks were reduced later.

Ethics did not slow them down.

It helped them scale safely.

III. Operationalizing the Advantage: Breaking Silos

A successful pilot in engineering means nothing if legal, sales, and operations are misaligned.

Scaling AI required structural integration.

5. Continuous Learning at Scale

For Microsoft, launching AI was not the end.

It was the start of learning.

Deployment was not a finish line.

It was the beginning of a data cycle.

When people used the tools in real work, that usage created new insight. Real-world behavior showed what worked, what failed, and what needed improvement. That data was used to retrain and refine the models.

At the same time, employees were not left behind. Microsoft invested in role-specific AI literacy training so people could understand how to use the tools confidently in their own jobs.

Legal and compliance teams were also part of the loop. Their feedback did not come after the release. It helped shape product roadmaps as systems evolved.

The result was simple but powerful.

Humans learned.

AI learned.

And both improved together.

6. Aligning Success Metrics

AI success at Microsoft was not judged by how advanced the technology looked.

It was judged by real business results.

Did developers write code faster?

Did Azure AI usage and revenue grow?

Were customers more satisfied?

Did support teams solve problems more quickly?

These were the measures that mattered.

Success metrics were shared across departments. Engineering, legal, product, and sales were not chasing different goals. They were aligned around common value outcomes.

Because everyone worked toward the same business impact, teams did not pull in different directions.

This alignment kept the organization focused.

And it prevented fragmentation as AI scaled.

IV. The Result: Intelligence Embedded, Not Installed

Today, AI at Microsoft is not a standalone initiative.

It is embedded across the ecosystem:

- Azure AI powers enterprise infrastructure globally.
- Copilot integrates into daily workflows for millions.
- GitHub has redefined how software is built.

AI became infrastructure — not an experiment.

The Core Insight

Microsoft did not succeed because its AI was flawless.

It succeeded because:

- Leadership modeled learning.
- Purpose-guided deployment.
- Humans remained accountable.

- Ethics scaled alongside speed.
- Cross-functional alignment replaced silos.

Technology scaled because judgment scaled.

That is the real blueprint.

AI transformation is not about installing intelligence.

It is about maturing the organization fast enough to handle it.

In that sense, Microsoft's real achievement was not Copilot.

It was leadership evolution.

Part IX

The AI-SENSE² Leader

Introduction

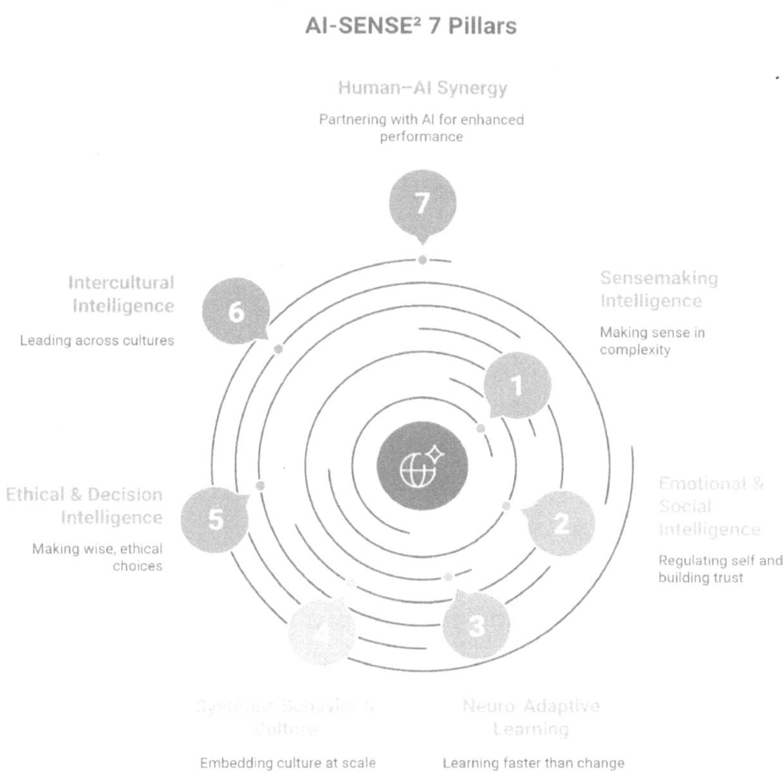

Throughout this book, we have explored the capabilities leaders need to thrive in an AI-accelerated world—sensemaking, emotional intelligence, learning agility, systemic culture, ethical judgment, and human–AI partnership.

But capabilities alone do not transform leadership.

They must eventually converge into something deeper: a leadership identity.

Part IX—**The AI-SENSE² Leader** marks that transition. This final section is about integration—bringing together the ideas explored throughout the book and showing how they operate as a unified way of thinking, deciding, and leading.

As artificial intelligence evolves toward **agentic systems**—systems capable of acting autonomously, coordinating across tools, and executing decisions at scale—the role of leadership shifts once again. The central question is no longer simply *how to adopt AI responsibly*. A more fundamental question emerges:

Who must leaders become in a world where intelligence itself is distributed across humans and machines?

This section introduces **AI-SENSE²** as a leadership operating system. Rather than a framework leaders apply occasionally, it becomes a continuous way of perceiving complexity, making decisions, learning from change, and guiding organizations forward.

Readers will explore why automation alone does not create intelligence—and why many early attempts at agentic AI struggle to deliver meaningful value. Often, the limitation is not technological capability, but the absence of innovation, agility, ethical clarity, and leadership readiness.

Without human sensemaking, learning velocity, and principled judgment, intelligent systems can amplify fragmentation rather than capability.

Through a lived leadership story, readers witness what it truly means to become **AI-ready** in practice. They see how leaders evolve their identity, shift from control to orchestration, and guide organizations through uncertainty while preserving trust, purpose, and human meaning.

The section concludes by looking toward the years ahead—2026 and beyond—offering a leadership playbook for a world defined by continuous change, intelligent systems, and growing complexity. Instead of attempting to predict specific technologies, it prepares leaders with the mindset and capabilities required to adapt to whatever emerges next.

By the end of Part IX, leadership is no longer defined by authority, expertise, or speed alone.

It is defined by the ability to integrate intelligence with wisdom, action with ethics, and innovation with humanity.

Outcome

Readers emerge with a **future-ready leadership identity**—able to sense, learn, decide, and lead alongside AI, using **AI-SENSE²** as their operating system for navigating the uncertainties and opportunities of the intelligent age.

Story: 36

From Automation to Intelligence: Preparing Organizations for the AI-Agent Era with AI-SENSE²

We are standing at a quiet but powerful turning point.

This moment is no longer about automating tasks, accelerating workflows, or squeezing efficiency from existing systems.

It is about **redefining how work itself happens**—with AI agents stepping in not as tools, but as **intelligent collaborators**.

Organizations that succeed in this next phase will not be the ones with the most advanced agents.

They will be the ones who **prepare human leadership, cognition, culture, and ethics** to work *with* intelligence at scale.

This is precisely what **AI-SENSE²** was designed to enable.

Sensemaking Intelligence — From Process Automation to Meaning Creation

Traditional automation followed rules.

AI agents follow **intent**.

They interpret ambiguity, adapt to changing conditions, learn from context, and know when to escalate to humans. This fundamentally changes leadership's role.

The key organizational shift is no longer:

"What process can we automate?"

But instead:

"What decisions can humans and AI make better together?"

This is **Sensemaking Intelligence** in action—helping leaders and teams interpret complexity, integrate multiple perspectives, and arrive at shared understanding before action.

Without sensemaking, AI speed only amplifies confusion.

Emotional & Social Intelligence — Trust Before Throughput

AI agents can generate answers instantly.

They cannot calm anxiety, resolve tension, or build trust.

As AI accelerates work, **emotional signals intensify**:

- Fear of loss of relevance
- Overwhelm from constant outputs
- Tension between human judgment and machine recommendations

AI-SENSE^2 places **Emotional & Social Intelligence** at the core—teaching leaders to regulate themselves, read team dynamics, and create psychological safety.

Trust becomes the invisible infrastructure that allows AI collaboration to scale without resistance.

Neuro-Adaptive Learning — Learning Faster Than the Technology Evolves

AI agents evolve weekly.

Human learning systems were never designed for that pace.

Neuro-Adaptive Learning closes this gap.

Instead of annual training or static onboarding, learning becomes:

- Embedded in daily workflows
- Triggered at the moment of need
- Reinforced through reflection and feedback

Leaders learn *how to learn again*—mirroring AI's ability to iterate, adapt, and transfer knowledge across contexts.

This is how organizations avoid falling behind their own tools.

Systemic Behavior & Culture — Designing Work That Actually Sticks

AI agents don't just change tasks.

They change **behavior patterns**.

Work no longer flows linearly. Multiple agents can operate in parallel—one clarifying requirements, another proposing options, another testing edge cases—while humans focus on judgment, creativity, and risk.

To make this sustainable, organizations must redesign:

- Roles (from executors to directors and reviewers)
- Rituals (from status updates to sensemaking checkpoints)
- Norms (from activity to impact)

AI-SENSE² treats culture as a **system**, not a side effect—embedding new habits so transformation sticks beyond pilots.

Ethical & Decision Intelligence — Wisdom Over Velocity

AI agents work fast.

Organizations must ensure they work **wisely**.

Ethical & Decision Intelligence ensures leaders:

- Define clear decision boundaries
- Establish human accountability
- Balance efficiency with values
- Override AI recommendations when context or ethics demand it

Trust collapses when AI decisions feel opaque or irresponsible.

It grows when transparency, explainability, and moral clarity are designed in from the start.

This is not governance for compliance—it is **decision intelligence for leadership**.

Intercultural × Human-AI Intelligence — Collaboration at Scale

AI agents don't sleep, don't localize naturally, and don't understand culture unless taught.

As organizations span geographies and time zones, leaders must orchestrate:

- Human–AI teaming models
- Intercultural collaboration norms
- Global alignment without central control

In AI-SENSE², the "²" represents this expanded intelligence—where **humans and machines collaborate across cultures, systems, and contexts**.

Workflows dynamically to whoever—human or agent—is best suited for the moment.

From Speed to Sustainable Performance

AI removes friction.

Humans must protect focus.

With administrative load reduced, leaders can create conditions for:

- Deep work

- Strategic thinking
- Creative synthesis
- Collective flow

But without attention management and energy awareness, AI simply increases exhaustion.

AI-SENSE² ensures that **performance is amplified without burnout**, aligning intelligence with sustainable human rhythms.

The Organization of the Near Future

In the near future:

- Humans set direction
- AI agents explore possibilities
- Both learn, adjust, and evolve together

Organizations that prepare today—by strengthening sensemaking, emotional regulation, learning agility, culture, ethics, and collaboration—will not just *adopt* AI agents.

They will **redefine meaningful work**.

The Core Truth

AI transformation is not a technology journey.

It is a **leadership evolution**.

AI-SENSE² is the operating system that makes this evolution possible—

transforming individuals into intelligent teams, and organizations into adaptive, globally resilient systems.

When humans lead with clarity, learning, ethics, and connection,

AI does not replace work.

It **elevates it**.

Story: 37

Preparing for the Future with AI Agility: The New Playbook for 2026 and Beyond

The future of work is arriving faster than most of us expected.

But it's not coming as a trend.

It's coming as a teammate—one that never sleeps, never forgets, and can execute faster than entire teams combined.

Agentic AI is about to transform the way we build products, lead teams, solve problems, and create value.

The question isn't *"Is AI coming?"* The question is *"Are we reorganizing ourselves for a world where humans think, and AI executes?"*

This is where **AI Agility** becomes the next superpower.

1. The Shift from Doing to Directing — [Cognitive Flexibility] + [Decision Intelligence]

In traditional teams, productivity meant speed and effort. In AI-enabled teams, productivity means **clarity and direction**.

AI will take over:

- documentation
- triage
- coordination
- test automation
- risk alerts
- optimization
- knowledge tracking

Your value shifts to something deeper: **How clearly you define the problem and how wisely you shape the outcome.**

This requires **Cognitive Flexibility** (shifting perspectives, reframing tasks) and **Decision Intelligence** (guiding AI with precise judgment).

The future belongs to the people who can guide AI with precision.

2. Build the Human–AI Operating Model — [Cognitive Flexibility] + [Learning Velocity]

Most organizations will treat AI as a tool. But genuine transformation happens when you **redesign the operating model**:

AI handles → execution, monitoring, automation, repetition. Humans handle → judgment, creativity, ethics, architecture, and leadership

This shift demands:

- **Cognitive Flexibility** to rethink roles and workflows
- **Learning Velocity** to absorb new ways of working quickly

This is not a replacement—it's elevation. We move from task ownership to insight ownership.

3. Let AI Carry the Cognitive Load — [Focus & Flow]

The greatest gift of AI is not speed. It is mental space.

When artificial intelligence takes care of reminders, documentation, follow-ups, scheduling, and routine checks, something powerful happens. The constant background noise that quietly consumes attention begins to fade. The small administrative tasks that fragment focus are handled in the background. Mental clutter reduces.

And when clutter reduces, clarity expands.

Instead of reacting to notifications and chasing minor tasks, the mind becomes available for higher-level thinking. You begin to see patterns instead of isolated problems. You think more strategically about direction, not just execution. You imagine possibilities that were previously crowded out by operational noise. You create with depth instead of rushing through surface-level output.

This is the true essence of Focus and Flow. AI becomes a silent cognitive partner that manages the mechanics of work so humans can access their best thinking. It protects attention, preserves creative energy, and opens space for insight.

Speed may increase productivity.

But mental space transforms performance.

AI agility means giving your brain the freedom it needs to do world-class work.

4. Outcome Thinking Becomes the New Currency — [Decision Intelligence]

In an AI-first workplace:

"I completed this task" no longer matters.

Instead, leaders will ask:

- What outcome did you shape?
- What judgment did you apply?
- What decision did you influence?

This is pure **Decision Intelligence** — blending intuition, values, data, and foresight.

We move from effort to **impact**.

5. Strengthen the Skills AI Cannot Replace — [Emotional Adaptability] + [Cognitive Flexibility] + [Decision Intelligence]

To thrive in the agentic era, invest in the un-automatable:

- **Judgment** — navigating ambiguity
- **Systems thinking** — seeing interdependencies
- **Creativity** — imagining the unseen
- **Communication** — aligning minds
- **Ethics** — ensuring responsibility
- **Empathy** — connecting with people and purpose

These reflect:

- **Emotional Adaptability** (empathy, composure)

- **Decision Intelligence** (ethical reasoning)
- **Cognitive Flexibility** (creative reframing)

AI enhances your abilities. But your humanity amplifies your leadership.

6. Embrace an Async-First Culture — [Focus & Flow]

With AI maintaining continuity, the old rhythm of work collapses.

AI agility means: ✓ fewer meetings ✓ fewer interruptions ✓ fewer status updates ✓ more deep work ✓ more focused collaboration ✓ more meaningful conversations

This supports **Focus & Flow** — uninterrupted, high-impact mental states.

Sync for connection. Async for execution. AI for flow.

7. Don't Retrofit — Redesign — [Cognitive Flexibility] + [Learning Velocity]

The biggest mistake teams will make is trying to force AI into old ways of working.

Instead, ask:

"If AI had existed when we built this workflow, what would we remove, reinvent, or reimagine?"

This mindset requires:

- **Cognitive Flexibility** to break patterns
- **Learning Velocity** to rebuild quickly

The future rewards teams that redesign the way work works.

Why AI Agility Matters Now

Because the future won't reward faster hands. It will reward sharper minds.

AI handles execution. Humans provide:

- clarity (**Decision Intelligence**)
- adaptability (**Emotional Adaptability**)
- creativity (**Cognitive Flexibility**)

- resilience (**Emotional Adaptability**)
- continuous growth (**Learning Velocity**)
- deep work (**Focus & Flow**)

If we prepare now, 2026 won't feel like disruption. It will feel like opportunity.

AI Agility isn't a skill. It's the new foundation for personal mastery, team performance, and organizational transformation.

AI-SENSE^2: 7 Pillars

Pillar	Description
Sensemaking Intelligence	Navigate complexity, interpret information
Emotional & Social Intelligence	Regulate emotions, build trust
Neuro-Adaptive Learning	Accelerate skill acquisition, adapt quickly
Systemic Behavior & Culture	Engineer organizational norms, drive change
Ethical & Decision Intelligence	Make values-aligned decisions
Intercultural × Human-AI Intelligence	Lead across cultures, collaborate with AI
Foundation – Human + AI Synergy	Align human capabilities with AI collaboration

Daily Actions to Become Fully Ready for Future AI Agility

Below is a **"Daily 20-Minute AI Agility Protocol"** every team member can follow.

1. Morning Mental Reset (3 minutes)

Purpose: strengthen *Emotional Adaptability* + *Focus & Flow*

Action:

- Start the day with 3 deep breaths
- Ask: **"What mental state do I want to bring into today?"**
- Set one emotional intention: calm, curious, courageous, open, focused.

Why: You cannot guide AI well if your mind begins in chaos.

2. Clarity Question of the Day (2 minutes)

Purpose: build *Decision Intelligence*

Ask yourself each morning:

"What outcome do I need, not what tasks do I need to do?"

Write down **1 clear outcome** you want to shape today.

Why: AI executes tasks. Humans shape outcomes.

3. Micro-Learning Burst (5 minutes)

Purpose: accelerate *Learning Velocity*

Choose **one** each day:

- read a short article
- watch a 3-minute explainer
- explore a new AI tool feature
- learn a leadership insight
- skim a case study

Why: 5 minutes daily = 150+ insights per month. Learning velocity compounds.

4. Daily AI Partnership Practice (5 minutes)

Purpose: build *Cognitive Flexibility + Human–AI collaboration*

Use AI (like ChatGPT or Copilot) for one small task:

- reframing a problem
- generating options
- analyzing risks
- simplifying instructions
- rewriting a message
- planning a meeting agenda

Your rule: Not doing → thinking. AI is doing → execution.

5. Outcome Review Loop (3 minutes) — End of Day

Purpose: improve *Reflection*, *Decision Intelligence,* and *Learning Velocity*

Ask:

1. **What did I learn today?**
2. **Where did AI help me think better?**
3. **What will I do differently tomorrow?**

This builds "adaptive intelligence."

6. Focus Sprint (Optional 20 minutes)

Purpose: strengthen *Focus & Flow*

Choose one complex task and work on it with:

- no meetings
- no Slack
- no notifications
- no context switching

Train your brain for deep work. Modern AI-powered environments reward humans who can focus deeply.

If your team can think clearly, learn rapidly, adapt emotionally, decide wisely, and stay in flow — AI will amplify their greatness, not replace their work.

Story: 38

Sensemaking Over Simulation: The Leadership Gap in Real-World AI

Building AI That Survives the Real World

When the engineering team from **Luminex Systems** arrived in the coastal city of **Varuna**, they were confident.

Their AI models were flawless in controlled environments. Accuracy scores were near perfect. Dashboards glowed with clean graphs and reassuring metrics. To them, this was just another deployment—another successful transition from lab to production.

Varuna had other plans.

By the end of the first week, the team faced a truth no technical manual had prepared them for:

AI doesn't fail because algorithms are weak.

AI fails because reality is complex.

Sensemaking Intelligence — When Reality Breaks the Model

Beneath Varuna's streets lay decades of undocumented history: rusted pipelines, abandoned fibre cables, forgotten infrastructure layered upon itself.

No accurate maps. No reliable drawings.

Every excavation uncovered surprises. Every network plan collapsed when it met the ground. Power fluctuated unpredictably. Connectivity vanished in certain zones without explanation. Night-time camera feeds dissolved into darkness as light levels dropped below usable thresholds.

The AI—trained on pristine datasets—was suddenly blind, starved, and confused.

The failure wasn't technical.

It was **a failure of sensemaking**.

The team had designed for an idealized world—not the one they were standing in.

Only when they stepped back did they realize:

AI is not built in data centers.

It is built in trenches, rain, heat, politics, constraints, and chaos.

Ethical & Decision Intelligence — Aligning AI with Reality

The turning point came when leadership reframed the mission.

Instead of forcing the environment to fit the AI, they aligned the AI to serve reality.

They conducted deep feasibility studies—mapping terrain like archaeologists, assessing power grids, and understanding environmental and social constraints. Privacy boundaries were clearly defined. Data collection was governed with transparency.

Ethics was not an afterthought. It was foundational.

Decisions were no longer about what *could* be automated, but what *should* be augmented—and where humans must remain accountable.

This shift transformed the AI from a brittle system into a responsible one.

Systemic Behavior & Culture — Engineering for Resilience

Design changed.

Where one network path failed, they added redundancy.

Where light was unreliable, they deployed infrared illumination.

Where power fluctuated, they built dual UPS systems.

Where connectivity dropped, they introduced edge processing.

The work was no longer about features.

It was about **systemic resilience**.

New rituals emerged:

- Field-first validation before deployments
- Cross-team reviews instead of siloed approvals
- Failure treated as signal, not blame

Culture adapted because behavior was redesigned—not enforced.

Neuro-Adaptive Learning — Learning Faster Than Change

Every failure became feedback.

Broken cables, blind spots, false alerts—none were hidden. All were studied.

The team adopted rapid learning loops:

- Deploy → observe → refine
- Fail → analyze → adapt
- Scale → monitor → evolve

The AI learned from cleaner, more honest data.

Humans learned to interpret AI outputs with sharper judgment.

Learning velocity became the competitive advantage.

Emotional & Social Intelligence — Regulating Pressure, Building Trust

The early weeks were intense.

Deadlines slipped. Pressure mounted. Field teams, data teams, and leadership didn't always agree.

But leaders anchored the work in emotional regulation.

They slowed reactions. They reframed setbacks. They created psychological safety when frustration peaked.

Calm leadership stabilized the system.

Trust replaced tension. Collaboration replaced defensiveness. Decisions improved because people were regulated, not reactive.

Intercultural × Human-AI Intelligence — One Integrated System

As the system matured, something subtle but powerful happened.

AI stopped being "the system."

It became a **teammate**.

AI handled pattern detection, diagnostics, and prediction.

Humans applied judgment, ethics, and context.

Field engineers, data scientists, operators, and leadership worked as one coordinated organism—across roles, disciplines, and technologies.

Workflows were redesigned so humans and AI complemented one another, not competed.

This was **a human–AI partnership at scale**.

Flow, Focus, and Sustainable Performance

With clarity restored, attention was protected.

Noise dropped. Signal increased.

Teams entered states of deep focus—solving real problems without drowning in alerts or dashboards. Energy stabilized. Burnout declined. Performance rose without sacrificing well-being.

AI carried the cognitive load.

Humans carried the responsibility.

The Outcome: AI That Thrives in the Real World

When the system finally scaled across Varuna, it didn't just function—it endured.

- Reliable performance across day and night
- Stable operations despite environmental uncertainty

- Ethical governance and public trust
- A resilient, learning organization capable of evolving

Varuna didn't adopt AI.

Varuna partnered with AI.

The Core Message

AI succeeds not because code is perfect—but because leaders are willing to face reality with clarity, ethics, learning, and humanity.

AI-SENSE² is not a technology framework.

It is a **leadership operating system** for building AI that works beyond the lab.

When humans evolve their thinking, behavior, and collaboration,

AI evolves with them.

That is how AI survives the real world.

Story: 39

From AI Experimentation to Enterprise Intelligence: The AI-SENSE² Transformation

> **"We are entering a new world in which data may be more important than software."**
>
> **— Sundar Pichai**

In the early days of the organization's artificial intelligence journey, excitement filled the halls of its technology teams. A global retail technology company had invested heavily in AI. Across departments, teams experimented with chatbots to improve customer service, recommendation engines to personalize shopping experiences, predictive analytics to forecast demand, and automation tools to streamline internal operations.

Individually, these experiments worked remarkably well.

Models produced accurate predictions. Chatbots answered questions quickly. Data scientists demonstrated elegant dashboards showing patterns hidden in vast streams of information.

Yet something puzzling began to emerge.

Despite millions of dollars invested in artificial intelligence, the company's overall performance had barely changed. Operational bottlenecks persisted. Decision cycles remained slow. Employees complained about navigating too many tools. Leaders found themselves overwhelmed by data yet uncertain about what actions to take.

AI seemed to exist everywhere—but meaningful impact remained elusive.

The realization slowly dawned across the leadership team: the challenge was not technological capability. The organization had strong engineers, talented data scientists, and access to powerful tools.

The problem was deeper.

The organization lacked a coherent way to **understand, trust, learn from, and govern AI.**

When AI Exists Without Alignment

As leaders began examining the situation more closely, several structural problems surfaced.

AI initiatives had grown organically across departments with little coordination. Different teams built models for similar problems without sharing insights. Business leaders often hesitated to rely on automated recommendations because they could not fully understand how those recommendations were generated. Engineers focused intensely on model accuracy but rarely examined how their systems fit into real workflows.

Data governance was inconsistent. Some teams carefully documented models and decisions; others relied on informal practices.

Perhaps most importantly, many employees quietly feared what AI might mean for their future roles. Even when leaders spoke about efficiency and innovation, the unspoken question remained: *Will automation replace me?*

In this environment, artificial intelligence began to resemble a collection of disconnected experiments rather than a coherent system of intelligence.

And this is where many organizations stall.

A Different Question

The turning point came during a leadership workshop examining the company's digital strategy. The conversation began with a familiar question: *Where else can we deploy AI?*

But one executive reframed the discussion.

Instead of asking where AI could be deployed, she asked a deeper question:

"How must our leadership, culture, and systems evolve to work intelligently with AI?"

That question changed everything.

The company adopted a new perspective grounded in the **AI-SENSE² leadership model**, recognizing that successful AI transformation requires far more than deploying algorithms. It requires leaders to develop new ways of understanding complexity, guiding teams, learning from data, redesigning systems, and balancing human judgment with machine intelligence.

AI-SENSE² in Action

The transformation began with **sensemaking intelligence**.

Leaders stepped back and mapped the organization's critical decision systems. Rather than focusing on isolated tools, they examined how information flowed across departments, where decisions were made, and where bottlenecks or confusion slowed progress.

This exercise revealed a surprising pattern. In many cases, data already existed—but the context required to interpret it was missing. AI could generate insights, but decision-makers lacked the broader understanding needed to act on those insights confidently.

Gradually, AI moved away from isolated tools toward **decision-support systems embedded directly into workflows**.

The second shift involved **emotional and social intelligence**.

Leadership realized that AI adoption would fail if employees felt threatened by automation. Instead of presenting AI as a replacement for human work, they reframed it as a powerful collaborator. Engineers, analysts, and operational teams were invited to participate in redesigning workflows.

These conversations transformed the atmosphere of the transformation. Fear began giving way to curiosity. Trust began replacing resistance.

The organization then focused on **neuro-learning intelligence**.

Rather than treating AI models as static tools, teams built continuous learning loops. AI systems captured outcomes and feedback from real-world decisions. Humans reviewed those results, corrected errors, and refined models over time.

Learning was no longer confined to training sessions or post-mortem meetings.

It became a **shared process between humans and machines.**

Next came **systemic behavior and culture**.

Leaders recognized that poorly designed processes become more dangerous when automated. Before scaling AI further, teams carefully mapped critical workflows. Bottlenecks were removed. Decision ownership was clarified. Handoffs between teams were simplified.

Only then was automation introduced.

AI stopped being an add-on to existing processes. Instead, it became a natural part of how work flowed through the organization.

Equally important was **ethical and decision intelligence**.

Leaders introduced governance structures that clarified when AI could act autonomously and when human approval remained essential. Systems were designed to explain recommendations, record decisions, and allow intervention when necessary.

Transparency replaced blind automation.

Finally, the organization embraced **human–AI synergy**.

AI excelled at detecting patterns across vast datasets and operating at scale. Humans continued to provide strategic judgment, ethical oversight, and creative problem-solving.

The relationship between people and technology shifted from competition to collaboration.

The Results

Within eighteen months, the impact became visible across the organization.

Operational decision speed improved significantly. Incident response times declined as AI surfaced relevant insights faster than manual analysis. Employees reported greater confidence in using AI systems because they understood how those systems worked and how they could intervene when necessary.

AI adoption expanded organically across departments.

Most importantly, artificial intelligence stopped being viewed as a technology initiative owned by the data science team.

It became a **core organizational capability.**

The Deeper Lesson

The company's transformation did not occur because it built the most sophisticated AI models.

It occurred because leadership aligned technology with **culture, systems, learning, and governance**.

Technology created possibilities.

Leadership created impact.

Reflection: The AI-SENSE² Principle

This story illustrates a central principle of the AI-SENSE² framework:

AI does not replace leadership.

It reveals the quality of leadership.

Organizations that integrate sensemaking, emotional intelligence, learning agility, systemic design, ethical judgment, and human–AI collaboration can transform AI into a lasting advantage.

Those who ignore these capabilities will struggle, no matter how advanced their technology becomes.

> **"The real danger is not that computers will begin to think like humans, but that humans will begin to think like computers."**
>
> **— Sydney J. Harris**

Key Lessons

Several lessons emerge from this transformation:

- AI success depends more on **organizational intelligence** than on model performance.
- Leaders must redesign **systems, culture, and decision structures** before scaling AI.
- Human judgment remains essential even in highly automated environments.
- The AI-SENSE^2 framework provides a practical blueprint for building **sustainable human–AI collaboration**.

In the age of intelligent machines, the organizations that thrive will not simply deploy better algorithms.

They will develop **better leadership.**

The AI-SENSE² Leadership Framework: Tools, Diagnostics, and Implementation

Introduction

> **"AI will not replace humans, but humans using AI will replace humans who don't."**
>
> **— Ginni Rometty**

As artificial intelligence reshapes organizations, many leaders focus on tools, models, and technical capability. Yet the real competitive advantage in the AI era will not come from technology alone—it will come from **how leaders think, decide, and guide intelligent systems within complex human environments.**

This is where **AI-SENSE²** becomes essential.

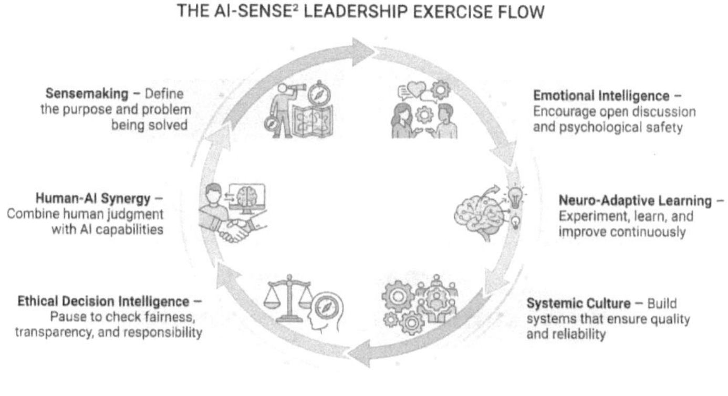

This section introduces AI-SENSE² as a **leadership operating system for an AI-driven world**—a structured way for leaders to interpret complexity, make responsible decisions, and translate artificial intelligence into meaningful organizational outcomes.

The chapters begin by exploring a foundational question: **what makes AI-SENSE² impossible to replace?** In a world where algorithms continuously evolve, the enduring advantage lies in human capabilities that machines cannot replicate—judgment, context awareness, ethical reasoning, and systemic thinking.

From there, readers are introduced to the **AI-SENSE² leadership framework**, which integrates these capabilities into a coherent model for modern leadership. But understanding the framework is only the beginning. The section then moves into practical execution, providing leaders with concrete tools to apply AI-SENSE² in real organizational environments.

Readers will explore:

- The **90-day leadership playbook and checklist** that help leaders translate the framework into immediate action
- **Leadership, team, and organizational readiness diagnostics** to assess current capabilities and identify gaps
- A **year-one implementation roadmap** for embedding AI-SENSE² across systems and teams
- The common **leadership traps** that prevent organizations from successfully integrating AI
- A practical **coaching methodology** for developing leaders, teams, and organizational capability at speed
- And a governance approach that helps **product managers operationalize AI-SENSE² within enterprise AI systems**

Together, these chapters transform AI-SENSE² from a conceptual framework into a **complete leadership system—one that connects strategy, behavior, culture, and governance in the age of intelligent machines.**

By the end of this section, readers will not only understand AI-SENSE²—they will have the **tools, diagnostics, and roadmap required to implement it across leaders, teams, and organizations.**

The goal is simple but powerful:

to help leaders move from **experimenting with AI to leading intelligently with it.**

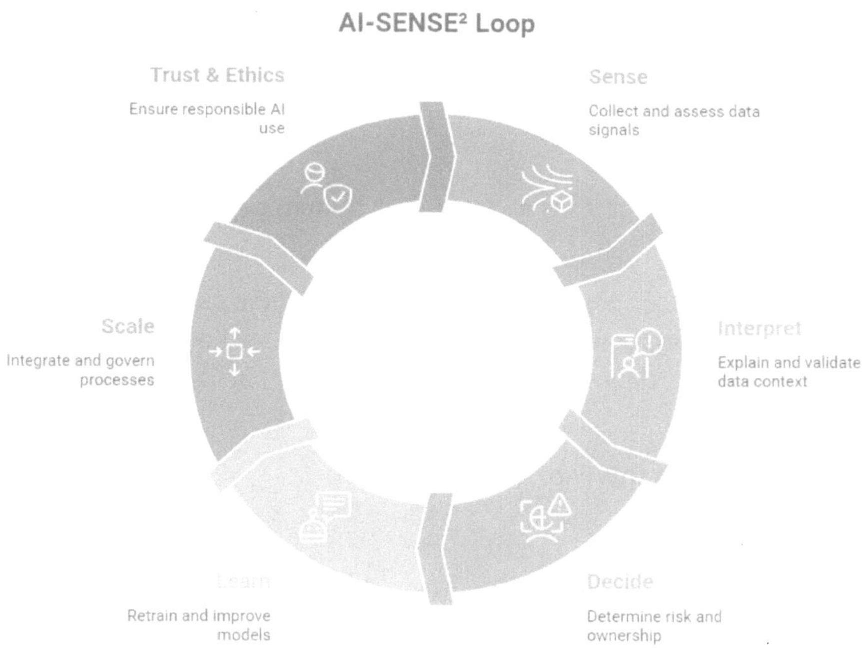

What Makes AI-SENSE² Impossible to Replace?

In a world flooded with AI frameworks, leadership models, and transformation playbooks, a fair question must be asked:

Why does AI-SENSE² matter?

More importantly, **what makes it impossible to replace?**

Is it simply another leadership framework adapted for the AI era?

Is it a rebranding of emotional intelligence with technology layered on top?

Or is it something fundamentally different?

To answer that, we must look at its core differentiators.

1. It Integrates Neuroscience with AI — Not Just Leadership Theory

Most AI frameworks focus on systems, tools, and data pipelines.

Most leadership frameworks focus on behavior, mindset, and culture.

Very few integrate the biology of human cognition with the reality of machine intelligence.

AI-SENSE^2 does.

It recognizes a simple truth:

AI is accelerating cognition at scale. But human cognition still governs judgment.

If leaders do not understand how attention works, how stress affects executive function, how bias distorts perception, or how neuroplasticity drives learning, they cannot lead effectively in an AI-accelerated environment.

AI-SENSE^2 bridges this gap.

It connects:

- Cognitive load and dashboard overload
- Emotional regulation and high-speed decisions
- Neuro-adaptive learning and continuous technological change
- Bias awareness and algorithmic governance

This integration of neuroscience and AI is not cosmetic. It is structural.

Because in the AI era, leadership failure is often neurological before it is strategic.

2. It Operates Across Three Layers: Self → Team → System

Many leadership programs focus on the individual.

Many transformation models focus on organizational systems.

Few explicitly connect the two in a measurable, scalable way.

AI-SENSE² is built on multi-layer development.

It strengthens:

- The individual leader's cognition and emotional regulation
- The team's psychological safety, collaboration, and shared meaning
- The enterprise's culture, decision loops, and governance

This layering matters.

Because AI initiatives fail when leaders grow individually, but systems resist change.

They also fail when systems are redesigned, but leaders lack the maturity to guide them.

AI-SENSE² ensures alignment across levels.

It is not self-development.

It is not corporate restructuring.

It is a synchronized transformation.

That architecture makes it difficult to substitute.

3. It Is Built on the Coherence Principle

Many organizations pursue intelligence.

Faster analytics.

Smarter models.

Better predictions.

But intelligence without coherence creates fragmentation.

Data teams optimize one metric.

Product teams optimize another.

Executives push for speed.

Legal teams slow for compliance.

Culture becomes strained.

AI-SENSE² is grounded in one unifying doctrine:

As intelligence increases, coherence must increase.

Coherence means alignment:

- Between strategy and execution
- Between speed and ethics
- Between human judgment and machine output
- Between innovation and accountability

Without coherence, AI scales confusion.

With coherence, AI scales clarity.

This principle turns AI-SENSE² from a skills model into a governance philosophy.

4. It Recognizes That Amplification Is the Real Risk

We are not entering the age of artificial intelligence.

We are entering the age of amplified intelligence.

Amplification does not fix weaknesses.

It magnifies them.

If leadership is reactive, AI accelerates reactivity.

If culture lacks trust, AI exposes fractures.

If decision-making lacks ethics, AI scales damage.

AI-SENSE² does not treat AI as a tool problem.

It treats it as an amplification problem.

That shift in framing is critical.

Because the real risk is not machines becoming powerful.

It is humans remaining unprepared.

Few frameworks confront this directly.

5. It Redefines Leadership as an Operating System

Most frameworks offer capabilities.

AI-SENSE² functions as a leadership operating system.

An operating system determines how everything interacts:

- How information flows
- How decisions are made
- How learning compounds
- How behavior becomes culture

AI-SENSE² integrates:

- Sensemaking under complexity
- Emotional and social regulation
- Neuro-adaptive learning
- Systemic culture shaping
- Ethical and decision intelligence
- Intercultural and human–AI synergy

These are not separate skills.

They are interdependent modules of a coherent leadership architecture.

Replacing one pillar weakens the whole.

That systemic integration makes the framework durable.

6. It Answers the Real Question Leaders Are Asking

Leaders today are not asking:

"How do I code better?"

They are asking:

- How do I stay relevant when AI learns faster than I do?
- How do I protect trust while accelerating innovation?

- How do I make wise decisions at machine speed?
- How do I scale AI without scaling risk?

AI-SENSE² addresses these questions holistically.

Not through technical instruction.

Not through motivational language.

But through structured, layered capability development.

The Core Differentiator, Sharpened

What makes AI-SENSE² impossible to replace is not one element.

It is the integration of:

- Neuroscience + AI collaboration
- Self → Team → System transformation
- A coherence-first doctrine
- An amplification-aware philosophy
- And a scalable leadership operating system

It does not treat AI as a tool.

It treats AI as a multiplier of human capability and human weakness.

And it prepares leaders accordingly.

Final Reflection

In the coming decade, organizations will not fail because they lack AI tools.

They will fail because intelligence outpaces coherence.

The leaders who thrive will not be those who know the most about algorithms.

They will be those who can align minds, culture, and machines into one coherent system.

That is the promise of AI-SENSE².

And that is why it cannot be easily replaced.

AI-SENSE²: The Leadership Operating System for an AI-Driven World

A New Kind of Leadership for a Fast-Changing World

The world is changing faster than ever.

New technologies appear every year. AI tools improve every month. Information moves in seconds. Decisions that once took weeks now need to be made in hours, sometimes minutes.

In the middle of all this change, many leaders feel the pressure.

The biggest challenge today is not learning how to code.

It is not building algorithms or understanding machine learning.

The real challenge is human.

How do you lead when technology moves faster than your strategy?

How do you stay relevant when AI seems to learn faster than you do?

How do you stay calm, think clearly, and make wise decisions when complexity keeps rising?

This is the age of **AI Agility**—an age where leadership depth matters more than technical skill.

It asks for a new kind of leader.

Not just a technologist.

Not a controller.

But a leader who can make sense of complexity, manage emotions, learn continuously, and work wisely with AI.

That is why **AI-SENSE²** exists.

AI-SENSE² is not a technology framework.

It is a **leadership operating system**.

It brings together human intelligence, neuroscience, ethical judgment, and AI collaboration to help leaders grow and stay grounded in a fast-changing, connected world.

The Leadership Crisis of Our Time

Today's leaders are pulled in many directions at once.

They face more data than ever before.

More speed.

More pressure.

More uncertainty.

AI-SENSE^2 does not treat these as separate problems.

It sees them as one connected leadership challenge.

Making Sense in a World Full of Noise

Leaders today are not lacking information.

They are overwhelmed by it.

Dashboards, reports, AI insights, opinions, and alerts arrive all the time. Instead of clarity, this often creates confusion. Decisions become reactive. Teams lose alignment.

Sensemaking intelligence brings leaders back to clarity.

It helps them slow down mentally, even when the world moves fast.

They learn to sit with uncertainty instead of panicking.

They connect different viewpoints into one shared understanding.

They learn to see patterns, not just problems.

AI can provide information.

But only leaders can create meaning.

Without sensemaking, AI increases confusion.

With sensemaking, AI supports wisdom.

Staying Calm and Building Trust

AI also increases emotional pressure.

There are constant alerts.

Faster decisions.

Public visibility.

And a quiet fear of falling behind.

When leaders do not manage their emotions, stress spreads quickly. Teams become defensive. Trust breaks down.

Emotional and social intelligence helps leaders stay centered.

They learn to manage stress instead of passing it on.

They read team emotions, even in high-pressure moments.

They create psychological safety across roles and cultures.

They lead with calm presence, not urgency or fear.

AI may increase speed.

Only emotionally balanced leaders protect trust.

Learning Faster Than Change

Many leaders carry a deep fear: becoming irrelevant.

AI learns all the time.

Humans often learn only in short bursts.

Neuro-adaptive learning helps close this gap.

Leaders learn how to learn continuously.

They practice unlearning old habits.

They stay curious instead of defensive.

They treat change as training, not a threat.

The future belongs to leaders who learn like AI—

but with reflection, purpose, and self-direction.

Making Change Stick

Many AI initiatives fail, not because the technology is bad, but because behavior does not change.

People return to old habits.

Systems reward old ways of working.

Systemic behavior and culture focuses on this problem.

Leaders learn to design simple rituals and norms that guide daily behavior.

They shape systems, so new habits become easy and natural.

They reinforce learning across teams, not just individuals.

Culture is not what leaders say.

It is what systems make easy and safe to do.

Making Wise Decisions at High Speed

AI brings power and speed—but also ethical questions.

Bias, privacy, trust, accountability, and transparency are no longer technical topics. They are leadership responsibilities.

Ethical and decision intelligence helps leaders slow down just enough to choose wisely.

They learn to balance AI insights with human judgment.

They make decisions guided by values, not just efficiency.

They create clear guardrails and shared accountability.

AI can suggest.

Leaders must decide.

Leading Across Cultures and Machines

The "^{2}" in AI-SENSE^2 represents expanded intelligence.

Today's leaders work across countries, cultures, time zones, and technologies. They must lead humans and machines together.

This capability helps leaders design healthy human-AI partnerships.

They create workflows where AI supports people, not replaces them.

They lead inclusive, global teams with clarity and respect.

The future is not human versus AI.

It is human plus AI, guided by culture and context.

Why AI-SENSE² Matters

Many frameworks focus on only one area—technology, leadership, or ethics.

AI-SENSE² brings them together.

It is human-centered.

Grounded in neuroscience.

Anchored in ethics.

Designed for real organizations.

Built to scale.

That is why it resonates with leaders across industries.

Because it answers the real question leaders are asking:

"How do I stay clear, relevant, and trusted in an AI-driven world?"

The Promise

AI-SENSE² is about future-proofing leadership, not just technology.

It helps leaders build clarity in chaos.

Stay emotionally strong.

Learn faster than disruption.

Work with AI without fear.

Make better decisions.

Build resilient organizations.

The leaders who thrive in the next decade will not be those who know the most about AI.

They will be those who can make sense, stay calm, keep learning, shape culture, decide wisely, and collaborate across humans and machines.

Because the biggest upgrade your organization needs

is not a new model—

It is a new leadership operating system.

A Final Question

If the future belongs to leaders who think deeply, feel steadily, learn quickly, and work wisely with intelligent machines.

AI-SENSE²: A Leadership Framework for an AI-Driven World

AI-SENSE² is not a technology framework.

It is a **leadership operating system** designed for a world where machines think faster, decisions scale wider, and the cost of poor judgment is amplified by intelligence itself.

At its core, AI-SENSE² recognizes a defining truth of our time:

as artificial intelligence accelerates, leadership must deepen.

Most AI initiatives fail not because algorithms are weak, but because human systems—leaders, teams, cultures, and decision norms—are not ready to guide intelligence responsibly. AI-SENSE² was created to close this gap. It integrates how leaders **sense, feel, learn, decide, behave, and scale intelligence**—without losing humanity in the process.

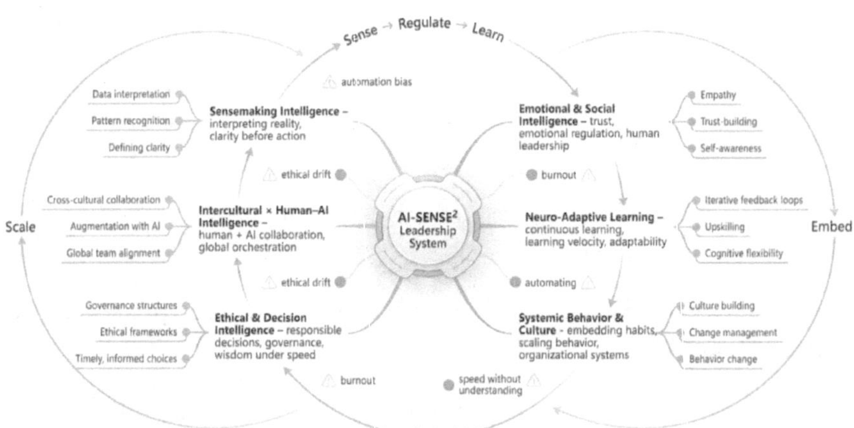

S — Sensemaking Intelligence

Seeing clearly before deciding

AI-SENSE² begins with **Sense** because leadership failures in the AI era rarely start with bad decisions—they start with **misinterpretation**. In environments flooded with data, dashboards, and predictions, leaders often react quickly while understanding less.

Sensemaking Intelligence enables leaders to pause, interpret, and frame reality before acting. It goes beyond perception and analysis to include awareness of context, consequence, and complexity—across both self and system. AI can surface patterns, but only humans can decide what those patterns mean.

Without Sense, speed becomes dangerous.

E — Emotional & Social Intelligence

Regulating self and leading humans through AI change

AI adoption is emotional before it is rational. Fear, identity threat, loss of control, and trust deficits emerge long before teams debate tools or models. AI-SENSE² places Emotional & Social Intelligence early because unregulated human systems sabotage even the best technology.

This pillar helps leaders regulate themselves under acceleration, build psychological safety, and maintain trust while change unfolds. As AI enters decision systems, leadership presence becomes louder—not quieter. How leaders show up emotionally determines whether AI empowers people or alienates them.

Without emotional regulation, intelligence amplifies instability.

N — Neuro-Adaptive Learning

Learning faster than change

In the AI era, expertise has a shrinking half-life. What leaders know today matters less than **how quickly they can learn, unlearn, and adapt** tomorrow. Neuro-Adaptive Learning shifts leadership from static mastery to learning velocity.

This pillar draws on neuroscience and real-world practice to show how leaders move from constant reaction to reflective presence, how they sustain cognitive flow while algorithms move fast, and how teams learn through feedback loops rather than linear training.

Without learning agility, leadership becomes obsolete—even when experienced.

S — Systemic Behavior & Culture

Making AI stick at scale

AI-SENSE² recognizes that intelligence does not scale through individuals—it scales through **systems**. Culture, incentives, decision rights, roles, and habits either reinforce intelligent behavior or quietly block it.

This pillar focuses on embedding AI-ready leadership into organizational behavior. It reframes coaching as transformation infrastructure, shows why AI fails at scale due to human friction, and explains how leaders redesign systems so judgment, alignment, and flow become everyday habits.

Without systemic alignment, AI remains stuck in pilots.

E — Ethical & Decision Intelligence

Wisdom under speed and scale

Accuracy is not the same as responsibility. As AI accelerates decisions, leaders must decide not only *what* AI can do—but *when it should pause.* Ethical & Decision Intelligence reasserts human accountability in AI-augmented environments.

This pillar treats ethics as a leadership capability, not a technical feature. It equips leaders to govern without slowing down, to balance speed with consequence, and to translate intelligence into trust—especially at the board and enterprise level.

Without ethical judgment, intelligence creates risk faster than value.

² — Intercultural × Human–AI Intelligence

Scaling intelligence across people, machines, and borders

The final dimension of AI-SENSE² acknowledges that modern intelligence is distributed across humans and machines, teams and agents, cultures and geographies. At this scale, leadership shifts from control to orchestration.

This pillar shows how AI-SENSE² functions as an enterprise and global operating system. It explains why agentic AI fails without innovation agility, how human–AI collaboration is operationalized responsibly, and how intelligence can scale without breaking trust, culture, or judgment.

Without this integration, intelligence fragments as it grows.

The Essence of AI-SENSE²

AI-SENSE² does not promise that leaders can control artificial intelligence.

Instead, it helps leaders take responsibility in a world where AI is everywhere and growing very fast.

It teaches leaders to pause and understand what is really happening before making decisions. It reminds them that people matter first. If people feel safe, calm, and respected, then systems work better.

AI-SENSE² also helps leaders learn quickly. Since the world is changing fast, leaders must learn faster than the change around them. This keeps them relevant and confident.

It encourages leaders to use AI in a healthy way — not just as a tool, but as part of the company's culture. It helps leaders move quickly without forgetting their values or good judgment. Even when working at a global level, it reminds them to keep the human connection strong.

With AI-SENSE², leadership changes. Instead of reacting quickly to everything, leaders think carefully before acting. Instead of relying only on expertise, they focus on learning fast. And instead of letting automation take over, they balance technology with wisdom.

In One Line

AI-SENSE² is a human-centered leadership framework that ensures AI accelerates judgment, not mistakes—and intelligence scales without losing meaning, trust, or humanity.

The 90-Day AI-SENSE² Leadership Playbook

How Leaders Adapt When Machines Think Faster

Why 90 Days Matter More Than the Next 3 Years

AI is not waiting for strategy cycles, operating models, or leadership readiness.

The leadership gap created by AI widens **every quarter**, not every decade.

This playbook exists because:

- AI adoption is accelerating
- Leadership capability is lagging
- The cost of delay is silent but compounding

The goal of the next 90 days is not AI mastery.

It is **leadership recalibration**.

When machines think faster, leaders must sense deeper.

The AI-SENSE² Leadership Rhythm

AI-SENSE² is not a checklist.

It is a **leadership rhythm** you practice repeatedly.

Sense → Regulate → Learn → Embed → Decide → Scale

If any step is skipped:

- Speed turns into risk
- Intelligence turns into noise
- Automation turns into abdication

This playbook sequences that rhythm into **three focused 30-day phases**.

Days 1–30: Sense Before You Decide

Leadership focus: Sensemaking Intelligence

In the first 30 days, the biggest risk is **reacting too fast**.

AI produces answers instantly—but leaders must:

- Frame the right questions
- Interpret context
- Understand second-order consequences

What leaders do differently

- Pause before approving AI-driven decisions
- Ask: *"What does this mean, not just what does it say?"*
- Separate signal from dashboard noise

Leadership shift

From speed → clarity

Days 1–30: Regulate the Human System

Leadership focus: Emotional & Social Intelligence

AI change triggers fear before logic:

- Fear of irrelevance
- Fear of loss of control
- Fear of invisible evaluation

Unregulated leaders amplify these fears.

What leaders do differently

- Replace certainty with curiosity
- Create psychological safety around AI use
- Acknowledge what they don't yet know

What to STOP in the first 30 days

- Forcing AI adoption
- Delegating judgment to tools
- Pretending confidence, you don't have

Leadership shift

From control → stability

Days 31–60: Learn Faster Than Change

Leadership focus: Neuro-Adaptive Learning

In the AI era, expertise expires quickly.

Learning velocity matters more than past success.

What leaders do differently

- Shift from "knowing" to learning loops
- Normalize experimentation and reflection
- Review decisions, not just outcomes

Simple practices

- Learning sprints (short cycles)
- Reflection after decisions
- Feedback before scale

Leadership shift

From expertise → adaptability

Days 31–60: Redesign System Behavior

Leadership focus: Systemic Culture

AI does not scale through individuals.

It scales through **systems**.

What leaders do differently

- Redesign decision rights
- Align incentives with judgment quality
- Use coaching as infrastructure, not support

Key realization

You don't scale AI.

You scale judgment, alignment, and trust.

Leadership shift

From hero leaders → healthy systems

Days 61–90: Decide When AI Should Pause

Leadership focus: Ethical & Decision Intelligence

Accuracy is not wisdom.

Confidence is not safety.

What leaders do differently

- Define when AI must escalate
- Pause AI without slowing the business
- Retain human accountability

Key questions leaders ask

- "Who owns this decision?"
- "What happens if this is wrong?"
- "Should speed yield to consequence here?"

Leadership shift

From accuracy → responsibility

Days 61–90: Turn AI Into Enterprise Trust

Leadership focus: Board-Ready Intelligence

At scale, AI must earn trust—not just deliver outputs.

What leaders do differently

- Translate AI into business value language

- Make risk visible, not hidden
- Govern without bureaucracy

Board-level mindset

AI strategy is a leadership confidence problem, not a data problem.

Leadership shift

From experimentation → enterprise confidence

Scaling Intelligence Without Losing Humanity

Leadership focus: Intercultural × Human–AI Intelligence

As AI agents and autonomy increase:

- Visibility decreases
- Decisions decentralize
- Cultural misalignment multiplies

What leaders do differently

- Orchestrate humans and machines
- Lead across cultures and algorithms
- Maintain shared judgment principles

Key insight

Scale without wisdom accelerates failure.

The AI-SENSE² Leader Identity

This playbook is not about becoming more technical.

It is about becoming **more human, more intentional, and more responsible**.

An AI-SENSE² leader:

- Senses before deciding
- Regulates emotion before scaling technology

- Learns faster than change
- Embeds intelligence into systems
- Governs wisely under speed
- Scales without losing humanity

The future does not belong to the fastest adopters.

It belongs to leaders who can guide intelligence with judgment and wisdom.

The 90-Day AI-SENSE² Leadership Checklist

A Practical Guide for Leading When Machines Think Faster

Purpose:

To help leaders recalibrate how they think, decide, and lead in the AI era—without overwhelm.

DAYS 1–30 | SENSE & EMOTIONAL REGULATION

Stabilize the leader before scaling intelligence

☐ **Sensemaking (Before Every Major AI-Influenced Decision)**

- ☐ I pause before acting on AI outputs
- ☐ I ask: *What does this mean? What might be missing?*
- ☐ I distinguish signal from dashboard noise
- ☐ I frame decisions in context, not just data

☐ **Emotional & Social Intelligence**

- ☐ I acknowledge uncertainty without projecting fear
- ☐ I replace certainty with curiosity in conversations
- ☐ I create psychological safety around AI experimentation
- ☐ I watch for automation bias—in myself first

☐ **STOP Doing in the First 30 Days**

- ☐ Stop chasing every AI trend
- ☐ Stop delegating judgment to tools
- ☐ Stop forcing AI adoption through authority

DAYS 31–60 | LEARNING & SYSTEM BEHAVIOR

Rewire how leaders and organizations adapt

☐ **Neuro-Adaptive Learning**

- ☐ I prioritize learning velocity over expertise
- ☐ I build short learning loops instead of long plans
- ☐ I review decisions, not just outcomes
- ☐ I normalize experimentation and reflection

☐ **Systemic Behavior & Culture**

- ☐ Decision ownership is clear
- ☐ Incentives reward judgment, not just speed
- ☐ Coaching is used as infrastructure, not support
- ☐ Alignment rituals exist across teams

☐ **Reality Check**

- ☐ AI pilots have a path to scale
- ☐ Human friction is addressed, not ignored

DAYS 61–90 | ETHICS & SCALE

Lead intelligence responsibly at the enterprise level

☐ **Ethical & Decision Intelligence**

- ☐ I know when AI should escalate or pause
- ☐ Human accountability is explicit

- ☐ Ethical consequences are discussed before execution
- ☐ Speed never overrides responsibility

☐ **Enterprise & Board Readiness**

- ☐ AI value is explained in business language
- ☐ Risks are visible, not hidden
- ☐ Governance enables speed, not bureaucracy
- ☐ Trust is actively built at senior levels

☐ **Scaling Without Losing Humanity**

- ☐ Human judgment remains central as autonomy increases
- ☐ Cultural and regional differences are respected
- ☐ Shared leadership principles guide AI use

FINAL SELF-CHECK (Day 90)

☐ I sense before deciding

☐ I regulate emotion before scaling technology

☐ I learn faster than change

☐ I embed intelligence into systems

☐ I govern wisely under speed

☐ I scale AI without losing humanity

If you can check most of these boxes, you are operating as an AI-SENSE² leader.

One-Line Leadership Reminder

You don't scale AI.

You scale judgment, trust, and leadership maturity.

AI-SENSE² Leadership Readiness Diagnostic

How prepared are you to lead humans, culture, and systems in the age of AI?

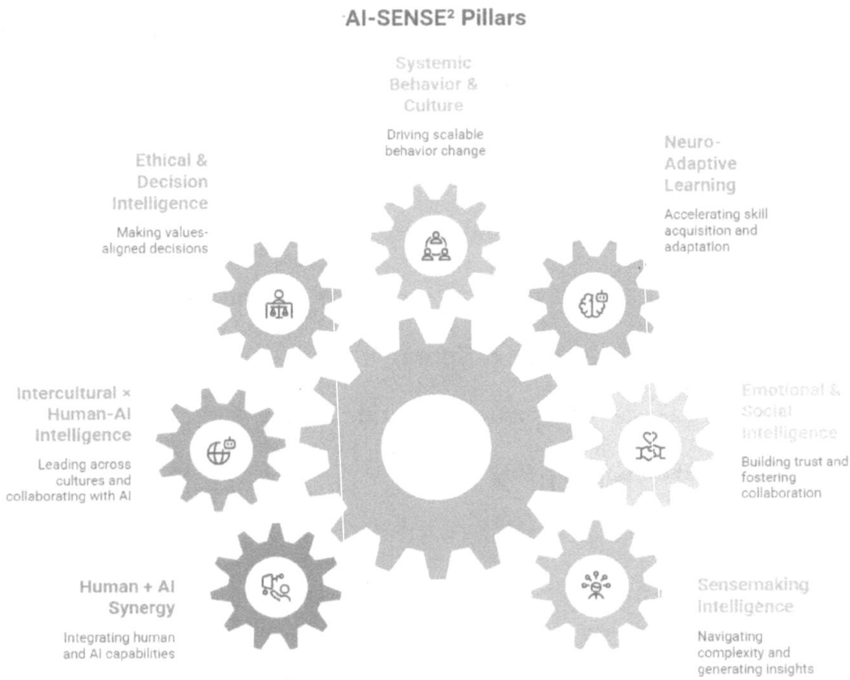

Instructions

For each statement, rate yourself honestly on a scale of **1–5**:

- **1 = Rarely true**
- **2 = Occasionally true**
- **3 = Sometimes true**
- **4 = Often true**
- **5 = Consistently true**

There are no "good" or "bad" scores—only signals.

I. Sensemaking Intelligence

Making meaning before making moves

Rate yourself:

1. I can clearly explain *why* we are using AI—not just *what* we are deploying.
2. When data conflicts, I slow down decisions to interpret meaning before acting.
3. I actively seek perspectives that challenge dominant narratives.
4. I can sense when something feels "off" even if metrics look strong.
5. I help teams align on shared understanding before execution begins.

Score (out of 25): ___

Reflection

- Where do I confuse more data with better understanding?
- What signals might I be overlooking because they don't fit existing assumptions?

II. Emotional & Social Intelligence

Regulating self, building trust, sustaining energy

Rate yourself:

6. I remain emotionally grounded under sustained uncertainty and pressure.
7. I openly acknowledge fear, ambiguity, or discomfort during AI-driven change.
8. I create psychological safety for people to question AI outputs.
9. I model calm curiosity rather than urgency or control.
10. I actively protect team energy, not just productivity.

Score (out of 25): ___

Reflection

- How does my emotional state influence decisions under pressure?
- Do people feel safe telling me when something doesn't feel right?

III. Neuro-Adaptive Learning

Learning faster than change

Rate yourself:

11. I regularly question habits that once made me successful.
12. I treat mistakes as learning signals, not competence failures.
13. I experiment in small cycles rather than waiting for certainty.
14. I unlearn outdated assumptions as deliberately as I learn new skills.
15. I build learning loops into daily work—not just training programs.

Score (out of 25): ___

Reflection

- What am I holding onto that may be limiting future relevance?
- How fast am I personally learning compared to how fast the environment is changing?

IV. Systemic Behavior & Culture

Embedding change so it sticks

Rate yourself:

16. I focus on changing systems and incentives—not just attitudes.
17. I notice when old behaviors resurface under pressure.
18. I redesign workflows to support human–AI collaboration.
19. I reinforce new behaviors consistently through rituals and norms.
20. I measure adoption and learning—not just outcomes.

Score (out of 25): ___

Reflection

- Where does our system quietly reward old behaviors?
- What behaviors would persist even if leadership attention disappeared?

V. Ethical & Decision Intelligence

Knowing when AI should pause

Rate yourself:

21. I am comfortable slowing down decisions when ethical uncertainty arises.
22. I explicitly define what AI should *never* decide alone.
23. I hold humans—not algorithms—accountable for outcomes.
24. I consider second- and third-order consequences of AI-driven decisions.
25. I ask who might be harmed, excluded, or disadvantaged by optimization.

Score (out of 25): ___

Reflection

- Where might speed be eroding wisdom?
- What decision today deserves more judgment than efficiency?

VI. Intercultural & Inclusion Intelligence

Leading across differences and scale

Rate yourself:

26. I adapt leadership style across cultures and contexts.
27. I recognize how AI can amplify cultural bias.
28. I design inclusion into systems—not just conversations.

29. I encourage diverse perspectives to challenge dominant thinking.
30. I balance global consistency with local nuance.

Score (out of 25): ___

Reflection

- Whose perspective is missing from our AI-driven decisions?
- Where might uniform systems be creating hidden friction?

VII. Human–AI Synergy

Thinking with AI without losing what makes us human

Rate yourself:

31. I use AI as a thinking partner—not an answer machine.
32. I know when to trust AI—and when to override it.
33. I delegate cognitive load without delegating responsibility.
34. I design workflows, where humans review judgment, not volume.
35. I protect deep thinking time from AI-driven noise.

Score (out of 25): ___

Reflection

- Where am I over-relying on AI—or under-using it?
- What human capability must never be automated?

Your AI-SENSE^2 Profile

Total Score (out of 175): ___

Interpretation Guide

- **140–175** → *Integrated AI-SENSE^2 Leader*

  You are actively building adaptive, ethical, and human-centered intelligence.

- **100–139** → *Developing AI-SENSE² Capability*

  Strong foundations, with gaps that could limit scale or sustainability.

- **70–99** → *AI-Exposed, Not AI-Ready*

  Technology may be ahead of leadership capability.

- **Below 70** → *High Risk Zone*

  AI adoption is likely creating invisible strain, resistance, or ethical risk.

One Final Question (The Most Important One)

Where am I still trying to lead with certainty—when the moment requires sensemaking, humility, and judgment?

AI-SENSE² is not about becoming faster than machines.

It is about becoming **wiser at human speed—at machine scale**.

AI-SENSE² Team & Organizational Readiness Diagnostic

How effectively are we operating as a human–AI system?

How to Use

- Have **each participant score independently**
- Then **compare averages + variance**
- High variance = misalignment risk (even if averages look good)

Scoring (1–5):

1 = Rarely true

2 = Occasionally true

3 = Sometimes true

4 = Often true

5 = Consistently true

I. Sensemaking Intelligence (Team & System)

Do we create shared meaning before acting?

1. We have a clear, shared understanding of **why** AI initiatives exist.
2. Cross-functional teams interpret data together before making decisions.
3. Conflicting signals are discussed—not ignored or overridden.
4. We regularly question assumptions behind AI-driven insights.
5. Leaders align narratives before execution begins.

Score (out of 25): ___

Team Signal to Watch

- If scores vary widely → teams are executing without shared meaning.

II. Emotional & Social Intelligence

Can we operate under pressure without fear, blame, or burnout?

6. People feel psychologically safe questioning AI outputs.
7. Fear of job loss or irrelevance is openly acknowledged—not suppressed.
8. Leaders regulate urgency rather than spreading anxiety.
9. Teams trust one another across functions and hierarchies.
10. Energy and well-being are treated as operational priorities.

Score (out of 25): ___

Team Signal to Watch

- High delivery + low safety = delayed failure.

III. Neuro-Adaptive Learning

Can we learn faster than change?

11. Teams experiment in small, fast learning cycles.
12. Failed pilots are reviewed openly and constructively.

13. Outdated processes are actively unlearned.
14. Learning is embedded in workflows—not limited to training programs.
15. Insights transfer across teams and regions quickly.

Score (out of 25): ___

Team Signal to Watch

- Long post-mortems, slow adaptation = low learning velocity.

IV. Systemic Behavior & Culture

Do our systems reinforce the behaviors we say we want?

16. Incentives support learning, judgment, and collaboration—not just output.
17. AI tools are embedded into daily workflows—not used on the side.
18. Old habits resurface less often under pressure.
19. New rituals reinforce human–AI collaboration.
20. Adoption is measured as seriously as performance.

Score (out of 25): ___

Team Signal to Watch

- Culture statements without behavior change = culture debt.

V. Ethical & Decision Intelligence

Do we know when AI should pause—and who decides?

21. Decision authority between humans and AI is explicit.
22. Ethical trade-offs are discussed before deployment—not after incidents.
23. Humans remain accountable for AI-driven outcomes.
24. Second-order consequences are considered.
25. Governance enables learning rather than blocking progress.

Score (out of 25): ___

Team Signal to Watch

- "The model said so" language = accountability erosion.

VI. Intercultural & Inclusion Intelligence

Can we scale intelligence across cultures, contexts, and geographies?

26. AI systems respect regional and cultural differences.
27. Diverse perspectives influence decisions early.
28. Inclusion is designed into workflows—not retrofitted.
29. Global teams collaborate effectively with shared clarity.
30. Local nuance is preserved without fragmenting strategy.

Score (out of 25): ___

Team Signal to Watch

- One-size-fits-all AI = silent resistance at scale.

VII. Human–AI Synergy

Are we thinking with AI, or outsourcing thinking to it?

31. AI is used to augment judgment, not replace it.
32. Teams know when to trust AI—and when to override it.
33. Humans focus on exceptions, ethics, and context.
34. AI reduces cognitive load instead of increasing noise.
35. Deep thinking time is protected.

Score (out of 25): ___

Team Signal to Watch

- Faster output + poorer decisions = false productivity.

Organizational AI-SENSE² Profile

Total Score (out of 175): ___

Interpretation

- **140–175** → *Adaptive Intelligent Organization*
- **110–139** → *Scaling with Friction*
- **80–109** → *AI-Exposed, Human-Unready*
- **Below 80** → *High Risk Transformation Zone*

The Alignment Test (Critical)

Ask as a group:

Where did our scores differ the most—and why?

Misalignment matters more than averages.

Three System-Level Questions to Close

- Where is AI moving faster than trust?
- Where is speed eroding judgment?
- Which behaviors must change for AI to truly work here?

What This Diagnostic Reveals

- Where AI initiatives will **stall**
- Where burnout or resistance will **surface**
- Where ethics and accountability may **break**
- Where leadership—not technology—is the constraint

The Leadership Rhythm (Memorize This)

Sense → Regulate → Learn → Embed → Decide → Scale

If any step is skipped, leadership breaks down.

The AI-SENSE² Framework at a Glance

S — Sensemaking Intelligence

Seeing clearly before deciding

What it means:

Leaders interpret reality before acting.

Without it:

More data → less clarity.

Key shift:

From reacting fast → understanding first.

E — Emotional & Social Intelligence

Regulating self and leading humans through AI change

What it means:

Leaders stabilize human systems under acceleration.

Without it:

Fear, resistance, and automation bias.

Key shift:

From pushing adoption → building trust.

N — Neuro-Adaptive Learning

Learning faster than change

What it means:

Leaders adapt continuously as expertise expires.

Without it:

Experience becomes a liability.

Key shift:

From expertise → learning velocity.

S — Systemic Behavior & Culture

Making AI stick at scale

What it means:

Leadership behavior becomes an organizational habit.

Without it:

Successful pilots, failed scale-ups.

Key shift:

From individual effort → system design.

E — Ethical & Decision Intelligence

Wisdom under speed and scale

What it means:

Leaders know when AI should act—and when it should pause.

Without it:

Accurate decisions, harmful outcomes.

Key shift:

From speed → responsibility.

² — Intercultural × Human–AI Intelligence

Scaling intelligence across people, machines, and borders

What it means:

Leadership orchestrates intelligence globally.

Without it:

Fragmentation and agentic failure.

Key shift:

From control → orchestration.

Simplify the rhythm like this:

- **Sense.**
- **Regulate.**
- **Learn.**

- **Embed.**
- **Decide.**
- **Scale.**

Remember:

"If you skip one step, the system breaks."

If you move fast without sensing, you make careless decisions.

If you don't regulate emotions, fear spreads.

If you don't learn, your expertise becomes outdated.

If you don't embed changes into systems, nothing scales.

If you don't govern wisely, speed becomes danger.

That's not philosophy.

That's how systems behave.

AI-SENSE²: Year 1 Implementation Roadmap

From Concept to Organizational Reality

AI-SENSE² is not meant to remain a theoretical leadership model. It is designed to become a living operating system embedded in how leaders think, teams collaborate, and enterprises make decisions.

The first year focuses on four progressive phases: building cultural stability, accelerating learning, redesigning systems, and scaling responsibly.

Phase 1 (Q1): Establish the Cultural Baseline

Theme: Sensemaking & Emotional Grounding

Goal: Move from AI anxiety to grounded curiosity.

Before scaling AI, the organization must stabilize its emotional and cognitive foundation.

Leadership Diagnostic

Begin with an audit of decision-making patterns. Where is the organization reacting to dashboards instead of interpreting context?

Where is speed replacing judgment? This diagnostic identifies gaps between data availability and sensemaking depth.

Psychological Safety Reset

Host structured conversations addressing unspoken fears around AI. Leaders openly discuss identity threats, job displacement concerns, and trust issues. The goal is not technical clarity—it is emotional regulation.

Introduce the "SENSE" Ritual

Embed a simple but powerful habit into executive meetings:

Before acting on new data, ask:

"What does this information not reveal about reality?"

This small pause begins shifting the culture from reaction to reflection.

Phase 2 (Q2): Build the Learning Engine

Theme: Neuro-Adaptive Learning

Goal: Replace static training with continuous learning velocity.

AI evolves monthly. Leadership capability must evolve just as fast.

Micro-Learning Architecture

Move away from long, one-time workshops. Introduce short, frequent AI Agility modules that build cumulative mastery over time.

Failure-as-Signal Forum

Create a structured monthly session where teams present AI experiments that failed—and what was learned from them. This normalizes experimentation and transforms setbacks into intelligence.

Socratic Prompting Mastery

Train leaders to use AI not just for productivity, but for cognitive challenge. They learn to use structured prompts to test their own assumptions, strategies, and blind spots.

The organization shifts from consuming AI tools to thinking with AI systems.

Phase 3 (Q3): Execute a Systemic Pilot

Theme: Behavior, Workflow, and Human–AI Synergy

Goal: Move one core workflow from experimental pilot to intelligent partnership.

Rather than scaling everywhere, choose one high-impact workflow.

Explicit Role Definition

Map the workflow clearly. Define where human judgment ("Gut") leads and where AI processing ("Grid") supports. Remove ambiguity in responsibility.

Ethical Checkpoints

Introduce mandatory review gates within the workflow. These checkpoints ensure AI outputs are evaluated through ethical and contextual judgment.

Resilience Stress Test

Simulate an AI failure scenario—such as inaccurate outputs or system outage. Evaluate whether human teams can maintain continuity and sound judgment without overdependence on automation.

This phase ensures AI enhances resilience rather than creating fragility.

Phase 4 (Q4): Scale Wisdom Across the Enterprise

Theme: Governance, Metrics, and Intercultural Expansion

Goal: Codify AI-SENSE^2 for enterprise-wide application.

With one workflow stabilized, the organization prepares for structured expansion.

Establish a Living Governance Council

Shift ethics from policy documents to active leadership practice. A cross-functional board regularly reviews AI's impact on trust, culture, and decision quality.

Redefine Success Metrics

Move beyond measuring automation rates. Begin tracking:

- Decision velocity with maintained quality
- Subject matter expert capacity uplift
- Reduction in low-value manual workload

Cross-Functional or Cross-Regional Expansion

Apply lessons learned from earlier phases to additional departments or geographies. Use early wins as the template for scaling responsibly.

Year 1 Success Indicators

By the end of the first year, success is measured not by the number of bots deployed, but by capability maturity.

- Faster adoption of new AI tools
- Stable human-in-the-loop oversight
- Increased trust in AI-supported decisions
- Significant reduction in repetitive manual tasks
- Strong psychological safety around experimentation

Most importantly, leaders demonstrate improved clarity under pressure and stronger judgment in AI-assisted decisions.

Final Message for the Board

Year one of AI-SENSE^2 is not about technological expansion.

It is about leadership evolution.

If implemented correctly, the organization will not merely use AI.

It will think with it.

And that distinction determines whether AI becomes a multiplier of confusion or a multiplier of capability.

The Leadership Traps AI-SENSE^2 Prevents

These are the hidden traps many leaders fall into as technology accelerates.

Speed without sense happens when decisions move faster than understanding. Teams act quickly, but without clarity, direction is lost, and mistakes multiply.

Automation bias creeps in when people start trusting systems more than their own judgment. What was meant to support thinking slowly begins to replace it.

The Brilliant Intern Trap appears when AI is treated like a genius assistant whose outputs are accepted without challenge. Leaders forget that intelligence still needs guidance, context, and responsibility.

Ethical drift sets in quietly. Small compromises feel harmless at first, but over time, values blur, accountability weakens, and trust erodes.

Pilot success, enterprise failure occurs when AI works well in small experiments but collapses at scale because culture, governance, and behavior never change.

And burnout disguised as productivity emerges when constant activity, alerts, and output look like progress, while people inside the system slowly run out of energy and meaning.

What an AI-SENSE^2 Leader Does Differently

A leader of the future does not rush to decide. They pause first, taking time to understand the situation, the signals, and the deeper meaning behind the data before choosing a direction.

They stay aware of their emotions, knowing that unregulated fear or urgency can spread quickly when technology moves fast. By staying calm and centered, they create stability while scaling innovation.

They learn continuously, adapting faster than change itself. Instead of clinging to past expertise, they stay curious, open, and willing to unlearn and grow.

They focus on designing systems that shape daily behavior, not just bold strategies written on slides. They know lasting change comes from how work actually happens.

They govern with wisdom, balancing speed and responsibility. Decisions are made quickly, but never carelessly, and accountability is always clear.

And as they work with AI, they remain deeply human—using technology to amplify judgment and creativity, not replace empathy, values, or purpose.

The Final Statement

An AI-SENSE² leader doesn't compete with machines.

They guide intelligence with judgment, wisdom, and humanity.

The AI-SENSE² Coaching Methodology: Developing Leaders, Teams, and Systems at Speed

How to Coach Yourself, Others, and the System in the Age of AI

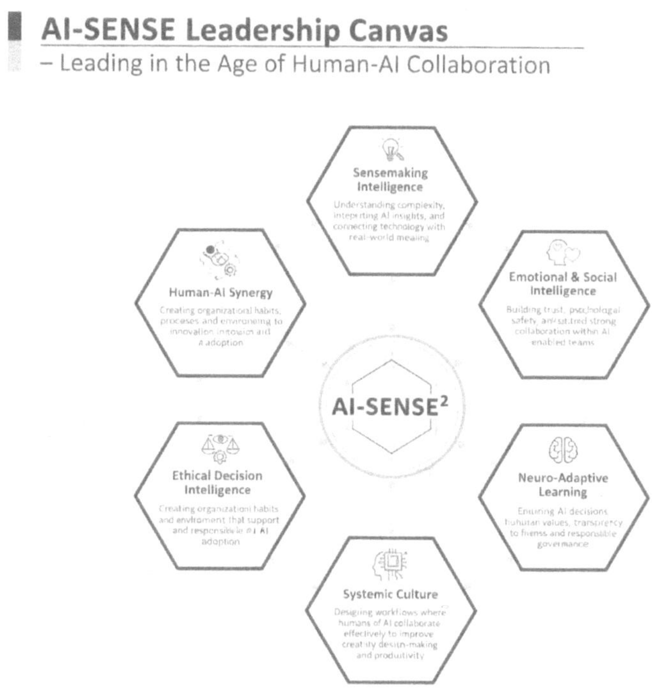

Leadership at AI Speed Requires a Different Kind of Skill

AI is changing how decisions are made, how work gets done, and how value is created—faster than any leadership model was designed to handle.

And this creates a quiet but dangerous gap.

Many leaders know *what* to do.

Few know *how to think, feel, and adapt* at the speed AI now demands.

That's why leadership without coaching begins to collapse under acceleration.

In the AI era, expertise alone doesn't scale. Authority doesn't hold. Control backfires.

What scales instead is the ability to **grow thinking—your own, your team's, and the system's—continuously**.

That ability is coaching.

Not coaching as a title.

Not coaching as a certification.

But coaching is a **leadership operating system**.

Why Coaching Becomes the Meta-Skill of the AI Age

In slower times, leaders could rely on experience and answers.

Today:

- AI evolves faster than any individual can keep up
- Knowledge expires quickly
- Emotional pressure increases as automation accelerates
- Teams face constant ambiguity instead of stable problems

In this environment, leadership based on answers becomes a bottleneck.

Coaching works because it does something different:

It doesn't provide answers—it **builds the capacity to find answers again and again**.

An AI-ready leader doesn't try to outthink AI.

They coach humans to **think better with it**.

A Shift Many Leaders Must Make

Rachel was respected, capable, and technically strong.

She understood AI, people, and complexity.

Yet something wasn't working.

Her team relied on her too heavily.

Decisions slowed.

Learning stayed individual instead of collective.

The turning point wasn't new tools or training.

It happened when Rachel stopped *demonstrating* good thinking and started **coaching it into others**.

She answered less.

She asked better questions.

She turned tension into reflection instead of resolution.

That's when the system changed.

Rachel didn't just become capable in AI-age leadership.

She became **generative**—able to grow that capability in others.

Coaching as a Three-Level Learning Flow

Effective AI-age coaching happens on three levels at the same time.

1. Self-Coaching: Inner Leadership First

Before leaders can coach others, they must first learn to **coach themselves**.

In a world where AI moves fast and decisions come quickly, self-coaching helps leaders stay steady instead of reactive. It allows them to pause, regulate their emotions, and respond thoughtfully rather than being driven by pressure or speed.

Through self-coaching, leaders notice important inner signals. They become aware when fear is shaping a decision, when ego is blocking learning, or when too much confidence is creating blind spots. They learn to shift consciously—sometimes thinking fast, sometimes slowing down for deeper reflection, and sometimes using AI as a thinking partner instead of a shortcut.

Instead of immediately asking, *"What should I do?"*, AI-ready leaders ask better questions.

They ask themselves what they are reacting to, what assumptions they may be making, and what new capability the situation is asking them to develop.

Over time, this way of thinking becomes a **daily habit**, not something used only during crises. It helps leaders grow steadily, make wiser decisions, and lead with clarity in an AI-driven world.

2. Team Coaching: Building Collective Intelligence

Teams don't really need more instructions or control.

What they need is better thinking together.

In the age of AI, team coaching is less about telling people what to do and more about helping them think clearly as a group. It focuses on making everyone's thinking visible, so assumptions and ideas are shared openly. It also helps teams notice emotions early—like stress, fear, or frustration—before these turn into conflict.

Instead of avoiding disagreement, good coaching helps teams turn friction into learning. AI insights are welcomed, but human judgment is not given up. AI supports the thinking—it does not replace it.

Rachel understood this well. She didn't add extra meetings or long workshops. Instead, she quietly built coaching into meetings the team already had. During planning sessions, she asked better questions. In decision reviews, she slowed the team down to examine assumptions. In retrospectives, she helped them reflect honestly. When AI outputs were discussed, she encouraged curiosity instead of blind trust.

Slowly, something changed.

The team stopped asking, *"What should we do?"*

They started asking, *"How should we think about this?"*

That shift—from rushing to execute toward learning how to think together—is the real move from execution to intelligence.

3. Systems Coaching: Changing Behavior at Scale

Most AI failures are not caused by individuals.

They happen because of the **systems** people work inside.

Systems coaching looks beyond single people and examines the environment around them. It pays attention to incentives that discourage learning, decision structures that create fear, governance rules that slow down good judgment, and cultural habits that make people afraid to speak up when they are unsure.

Rachel began to change her approach. Instead of asking who made a mistake, she started asking different questions. She looked at what behaviors the system was encouraging, what kind of thinking policies were rewarding, and what people felt unsafe saying out loud.

By coaching the system—not just the people—she discovered something important. Culture doesn't change through big announcements or loud interventions.

When leaders coach systems well, **culture changes quietly and permanently**.

Coaching Through the Seven AI-SENSE² Capabilities

Each capability requires a different coaching stance:

- **Sensemaking** → reframing complexity, connecting signals
- **Emotional & Social Agility** → regulation, empathy, psychological safety
- **Learning Agility** → curiosity, unlearning, faster adaptation
- **Systemic Behavior** → habits, rituals, reinforcement
- **Ethical Judgment** → values, consequences, accountability

- **Human-AI & Intercultural Intelligence** → inclusion and boundary clarity
- **Focus & Resilience** → energy, attention, sustainability

Leaders don't teach these as content.

They **coach them into existence** through reflection, questioning, and repetition.

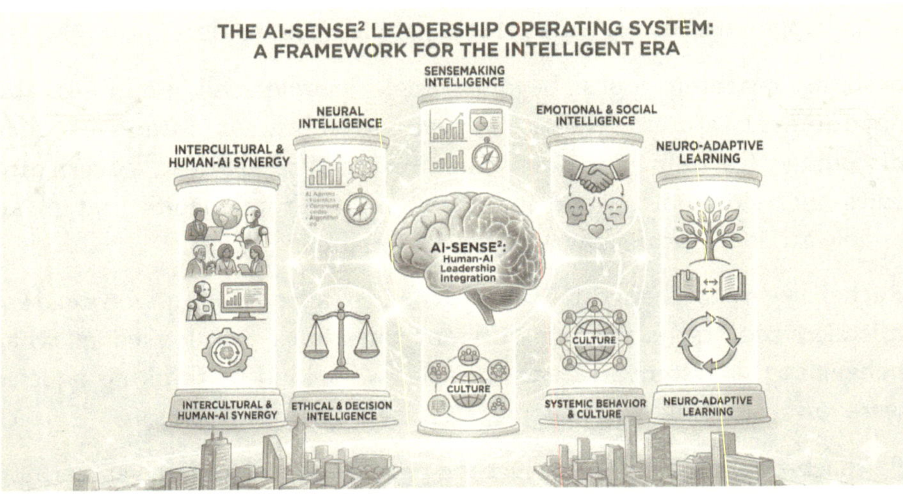

How Coaching Scales Without Burning Leaders Out

Rachel's biggest breakthrough wasn't coaching her team directly.

It was coaching her managers to coach their teams.

She introduced:

- Peer coaching circles
- Simple coaching conversation guides
- Shared reflection prompts after AI-driven decisions

Within months, coaching spread sideways and downward.

The organization didn't become "more coached."

It became **more thoughtful, resilient, and adaptive**.

What Makes This Approach Different

Coaching is often treated as:

- A soft skill
- A specialist role
- A separate activity

This approach makes a stronger claim:

In the AI era, coaching is **how leadership interfaces with intelligence**—human and machine.

It is:

- A daily leadership practice
- A system-shaping tool
- A scalable capability
- The bridge between AI insight and human wisdom

No jargon.

No certification barrier.

No abstract theory.

The Core Learning

AI doesn't need leaders who know more.

It needs leaders who can **grow more—faster—together**.

And that growth doesn't happen in workshops or frameworks alone.

It happens:

- in real decisions
- under real pressure
- inside real work

That is the essence of AI-SENSE^2 coaching:

not leading answers—but **building the capacity to think, decide, and adapt at speed**.

How leaders actually coach these capabilities into people and systems.

1. Sensemaking Intelligence

Coaching skills: Questioning, scenario facilitation, reframing complexity

Sensemaking coaching helps leaders and teams slow down *meaning*, not speed up answers. The coach's role is to surface assumptions, connect fragmented signals, and help people see the system rather than isolated data points.

Instead of asking, *"What does the data say?"* the coach asks:

- *What might this data be missing?*
- *What different stories could explain this pattern?*
- *What happens if this assumption is wrong?*

In an omnichannel retail environment, sensemaking coaching brought marketing, supply chain, and store teams into shared scenario discussions. Rather than optimizing each channel separately, teams explored customer journeys end-to-end. This reduced misalignment, cut rework, and accelerated decisions.

Outcome:

Faster alignment happens not because people agree faster—but because they understand each other sooner. Delays shrink when meaning is shared.

2. Emotional & Social Intelligence

Coaching skills: Active listening, emotional awareness, psychological safety

This pillar focuses on helping leaders notice and regulate emotional undercurrents—especially under pressure. Coaches here work less on *what* teams decide and more on *how* emotions shape those decisions.

The coach listens for:

- Fear disguised as resistance
- Overconfidence masking uncertainty
- Silence signaling psychological risk

During a high-pressure global product launch, emotional intelligence coaching helped leaders pause, acknowledge stress openly, and invite concerns without judgment. Instead of pushing harder, they created safety for the truth to surface early.

Outcome:

Engagement rises because people feel seen. Delivery improves because emotional friction no longer drains cognitive energy.

3. Neuro-Adaptive Learning

Coaching skills: Growth mindset cultivation, rapid feedback, unlearning

Neuro-adaptive coaching helps individuals and teams learn at the speed of change. The focus shifts from *getting it right* to *learning fast enough to stay relevant*.

Coaches reinforce:

- Experimentation over perfection
- Feedback as fuel, not failure
- Unlearning outdated habits as a skill

In an IT firm adopting AI tools, coaches embedded learning loops into daily work—short experiments, immediate feedback, and reflection. Learning stopped being an event and became a rhythm.

Outcome:

Adoption increases because people aren't waiting to feel "ready." Decisions speed up because learning replaces hesitation.

4. Systemic Behavior & Culture

Coaching skills: Behavior reinforcement, ritual design, culture shaping

This pillar recognizes that culture doesn't change through slogans—it changes through repeated behavior. Coaches here work at the system level, identifying which behaviors are rewarded, ignored, or punished.

The coaching focus is on questions like:

- *What behaviors does this system actually reinforce?*
- *What rituals shape daily decisions?*
- *Where do incentives contradict values?*

In sustainability initiatives, coaching helped teams embed simple rituals—visible tracking, peer accountability, and leadership modeling. Sustainability moved from aspiration to habit.

Outcome:

Adoption increases because behavior is supported by the system. Culture shifts because the environment reinforces the right actions daily.

5. Ethical & Decision Intelligence

Coaching skills: Ethical scenario exploration, bias awareness, judgment coaching

Ethical coaching develops leaders' ability to navigate gray areas—not by rules alone, but through deliberate judgment. Coaches help leaders recognize where optimization collides with values.

Instead of asking, *"Is this compliant?"* coaches ask:

- *Who could be harmed by this decision?*
- *What trade-off are we making explicit—or hiding?*
- *Would we stand by this decision publicly?*

In AI-driven lending, ethical coaching surfaced bias risks early. Leaders ran scenario reviews before deployment, not after complaints. Ethics became proactive, not reactive.

Outcome:

Trust increases because decisions feel fair and explainable. Complaints drop because issues are addressed upstream.

6. Intercultural & Inclusion Intelligence

Coaching skills: Inclusive facilitation, cultural intelligence, perspective-taking

This pillar equips leaders to work across cultures, geographies, and identities. Coaches help teams recognize that what feels "normal" in one context can feel exclusionary in another.

Coaching practices include:

- Slowing dominant voices
- Inviting non-obvious perspectives
- Surfacing cultural assumptions

In multicultural consulting teams, inclusive facilitation coaching reduced friction and miscommunication. Teams learned how cultural context shapes expectations around hierarchy, conflict, and decision-making.

Outcome:

Delivery accelerates because misunderstandings decrease. Attrition drops because people feel respected and included.

7. Human–AI Synergy

Coaching skills: Workflow design, augmented decision coaching, authority clarity

This pillar focuses on designing *how humans and AI think together*. Coaches help leaders clarify decision boundaries—what AI proposes, what humans decide, and where collaboration happens.

Key coaching questions include:

- *Where does AI add insight vs. where does it lack context?*
- *Who owns the final decision—and why?*
- *How do we prevent blind trust or blind rejection of AI?*

In logistics AI integration, coaching redesigned workflows so AI handled routing suggestions while humans reviewed exceptions. Accountability became clear, errors dropped, and efficiency rose.

Outcome:

Efficiency improves without sacrificing judgment. Trust grows because people know when to rely on AI—and when to override it.

AI-Enhanced Coaching Strategies

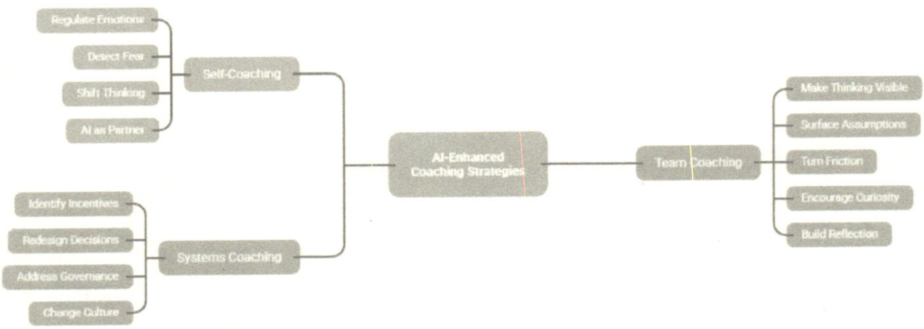

The Bigger Pattern

Across all seven pillars, one truth emerges:

Coaching is not about giving answers.

It is about building the capacity to think, decide, and adapt—again and again.

When leaders develop these coaching skills, AI stops being a fragile initiative and becomes a resilient capability—embedded in people, teams, and systems.

AI Governance for Product Managers: Operationalizing AI-SENSE² in Enterprise AI Systems

Artificial Intelligence amplifies intelligence.

But amplification does not automatically create wisdom.

If leadership is unclear, AI scales confusion.

If ethics are weak, AI accelerates harm.

If governance is absent, AI magnifies risk.

This is why AI governance is not a compliance activity.

It is a leadership discipline.

Through the lens of **AI-SENSE²**, AI governance becomes the structured way product managers ensure that as intelligence increases, coherence increases with it.

This article explains how AI-SENSE² translates into a practical AI Governance Framework for Product Managers building enterprise-ready AI systems.

AI-SENSE² as the Foundation of AI Governance

AI-SENSE² integrates:

Human leadership

Neuroscience

System thinking

Ethical intelligence

Human–AI collaboration

Governance is where these dimensions become operational.

Let us walk through governance using each pillar of AI-SENSE².

S — Sensemaking Intelligence

Governance Begins with Clarity

Before building any AI model, a product manager must first understand the complexity of the situation. Governance does not begin after the model is built. It begins with deep thinking and careful interpretation.

The first question is simple but powerful: What real business problem are we trying to solve? Then comes another important question: What decisions will this AI system influence? A model is never neutral. It shapes actions, priorities, and outcomes.

It is also necessary to examine the data.

What assumptions are hidden inside it?

Does it reflect past biases or incomplete realities?

Beyond data, there are broader forces to consider. Political pressures, regulatory requirements, and cultural expectations all influence how AI will be received and used. A responsible product manager also thinks ahead about unintended consequences.

What could go wrong?

Who might be affected in ways we did not expect?

Without this deep sensemaking, governance becomes paperwork. It turns into documents and approvals that exist only for compliance. But when sensemaking is strong, governance becomes strategic foresight. It helps the organization see risks and opportunities before they fully appear.

In practical terms, this means writing clear problem statements, so everyone understands the purpose of the system. It means defining how much risk the organization is willing to accept. It requires measurable success metrics so performance can be evaluated honestly. And it demands explicit decision boundaries, so it is clear where AI can act and where human judgment must take over.

AI systems should never operate in confusion or vague intent. Clarity is the first and most powerful mechanism of governance.

E — Emotional & Social Intelligence

Governance as Trust Architecture

AI systems do more than process data. They influence people.

Employees may worry about losing their roles. Customers may question whether decisions are fair. Executives may fear legal or reputational risk. Because of this, governance is not only a technical responsibility. It requires emotional maturity.

Product managers must create an environment where teams feel safe to experiment with AI and admit uncertainty. People should be able to question results without fear of blame. Risks should be discussed openly, not hidden. Collaboration across departments must be encouraged so that technology, legal, operations, and business leaders work together. Clear ownership should also be defined, so everyone knows who is accountable for decisions.

Governance often fails when teams hide doubts or avoid difficult conversations. When uncertainty is suppressed, risks grow quietly in the background.

Through strong Emotional and Social Intelligence, governance becomes relational rather than restrictive. It is built on trust, open dialogue, and shared responsibility. Over time, that trust becomes part of the product itself—embedded directly into how it is designed, deployed, and improved.

N — Neuro-Adaptive Learning

Governance as Continuous Evolution

AI systems are not static. They keep changing because the world around them keeps changing. Data evolves. Customer behavior shifts. Markets move. Over time, models begin to drift away from their original accuracy.

Because of this, governance cannot be fixed or rigid. It must learn and adapt faster than the changes happening around it.

Neuro-Adaptive Learning means product managers design systems that continuously watch for drift, schedule regular retraining, collect feedback from real users, and monitor performance over time. It also means carefully tracking versions and documenting updates so that every change is clear and traceable.

Governance is not just a checkpoint before launch. It is not something you approve once and forget.

It is a discipline that continues throughout the entire lifecycle of the AI system.

Without adaptive governance, an AI system slowly becomes fragile. Small shifts go unnoticed. Errors increase. Trust declines. Over time, the system that once created value can start creating risk.

S — Systemic Behavior & Culture

Embedding Governance into Organizational DNA

AI governance cannot rely only on individuals being careful. It cannot depend on one responsible manager or one alert engineer. If governance lives only in someone's personal discipline, it will eventually fail—especially under pressure.

Governance must be built into how teams work every day and how the organization operates as a whole.

This means every AI model should have a clearly defined owner. There should be regular checkpoints where performance, risk, and compliance are reviewed. Important decisions should involve cross-functional review boards so that technology, legal, clinical, and business perspectives are all considered. There must also be clear incident response processes, so everyone knows what to do if something goes wrong. Finally, enterprise-wide AI policies should guide behavior consistently across teams.

But processes alone are not enough. Culture determines whether governance truly survives when deadlines are tight and pressure increases.

If an organization rewards only speed and punishes caution, governance will slowly disappear. Teams will cut corners. Risks will be ignored. Short-term wins will replace long-term responsibility.

Through strong Systemic Behavior and Culture intelligence, product managers ensure that governance becomes part of daily operations. It becomes a habit built into workflows and decision-making—not an obstacle, and not an optional layer added at the end.

E — Ethical & Decision Intelligence

The Core of Responsible AI

At the center of AI governance is ethical judgment. Technology can move fast, but responsibility must move just as fast.

Product managers must make sure that bias is tested before deployment and fairness is monitored continuously after launch. Systems should be explainable so that decisions are not mysterious or impossible to understand. There must be clear ways for humans to override AI decisions when needed. Strong audit logs and traceability are essential so actions can be reviewed. Regulatory requirements should also be clearly mapped to the system's design and operation.

But ethics goes beyond policies and checklists.

It requires deeper reflection. Is this decision defensible if questioned publicly? Would we confidently stand behind this output in front of customers, regulators, or the media? Could this system unintentionally harm vulnerable groups? Are we optimizing for short-term profit while creating long-term risk?

Ethical and Decision Intelligence ensures that intelligence does not move ahead of responsibility. It keeps innovation aligned with values, protecting both people and the organization over time.

^{2} — Intercultural × Human–AI Intelligence

Governance at Global and Machine Scale

Enterprise AI systems do not operate in a single environment. They function across different countries, cultures, and regulatory systems. What works in one region may create risk or misunderstanding in another. Because of this, governance must take a global perspective.

Data laws differ from country to country. Some regions require data to remain local. Cultural norms also shape how people interpret automated decisions. A recommendation that feels neutral in one culture may feel biased or insensitive in another. Governance must ensure inclusion across diverse user groups so that no segment is unintentionally excluded or disadvantaged.

Product managers play a central role in defining collaboration between humans and AI. It must be clear where the system only recommends, where humans make the final decision, when escalation is required, and where full automation is appropriate. These boundaries should not be vague or assumed. They must be explicitly designed.

Human–AI synergy does not happen automatically. It must be carefully architected into workflows, policies, and responsibilities.

When collaboration boundaries are unclear, accountability weakens. People are unsure who is responsible for outcomes, and trust begins to erode. Clear structure protects both performance and responsibility.

AI Governance Across Self → Team → System

AI-SENSE² emphasizes multi-layer development.

AI governance must operate across:

Self

The product manager demonstrates:

- Ethical clarity
- Risk awareness
- Composure under pressure
- Courage to delay unsafe deployment

Team

Cross-functional collaboration ensures:

- Transparency
- Distributed accountability
- Shared understanding of risk

System

Enterprise structures enforce:

- Policy alignment
- Regulatory readiness
- Scalable governance mechanisms

When governance operates at all three levels, it becomes sustainable.

Common Governance Failures Through the AI-SENSE² Lens

When Sensemaking is weak → unclear problem definitions.

When Emotional Intelligence is absent → resistance and mistrust.

When Learning agility is low → unmanaged model drift.

When Systemic culture is fragile → governance shortcuts.

When Ethical intelligence is shallow → bias exposure.

When Human–AI clarity is missing → accountability confusion.

Governance failures are leadership failures.

AI-SENSE² addresses them structurally.

Why AI-SENSE² Strengthens Enterprise AI Proposals

Organizations evaluating AI vendors increasingly ask:

- How is bias monitored?
- How are decisions explained?
- How is governance structured?
- How does the system scale globally?
- What happens when the model fails?

An AI Governance Framework aligned to AI-SENSE² provides coherent answers.

It demonstrates that intelligence, culture, ethics, and systems are integrated.

This coherence builds executive confidence.

And confidence wins enterprise adoption.

The Governance Mindset Shift

Product managers must move beyond asking:

"How accurate is the model?"

They must ask:

"Is this system coherent, ethical, explainable, and defensible at scale?"

AI-SENSE² ensures intelligence rises with coherence.

Governance ensures coherence rises faster than intelligence.

Final Thought

AI governance is not about slowing innovation.

It is about stabilizing acceleration.

Through AI-SENSE², product managers become:

Architects of responsible intelligence

Designers of trust-based systems

Leaders who align humans, culture, and machines

In the age of amplified intelligence, governance is not optional.

It is the foundation of sustainable AI leadership.

Case studies

Case study 1

When AI Stops Talking and Starts Working: The Rise of Agentic AI

Today, many people talk about AI in a very shallow way. They argue about how big the AI model is, how good the prompt is, or how nicely AI can write, summarize, or search. These things sound impressive, but they are not the most important part.

When we look at real companies using AI well—for improving customer service or CRM systems moving beyond simple chatbots—we see something important. Their success did not come from better writing or smarter answers.

The real change was moving from **AI that gives answers** to **AI that takes action**.

Traditional AI waits for a question. It looks for information and gives a reply. Agentic AI works differently. It looks at a situation, makes a plan, and then completes a task step by step across different systems. This change makes AI much more useful.

Because of this, success should be measured differently. Many teams still ask, “Did the AI give a good reply?”

But smarter teams ask, "Did the job get done from start to finish?"

Real value comes when AI completes a task fully, without humans jumping in to connect systems manually.

Agentic AI works best in everyday boring work—not fancy labs or future demos. Think about daily tasks like reading an email, opening a ticket, copying data into another system, updating a customer record, and sending a reply. This is not thinking work. It is coordination work. And this is exactly what agentic AI is good at.

When teams clearly understand the full process—from start to finish—they know where an agent can help. Not a chatbot that talks, but an operator that works.

Many AI projects fail because teams treat agents like smarter chatbots. They give them more information, but no ability to act. An AI agent is only powerful if it can use real tools.

Instead of just answering "Where is my refund?", a good system lets the agent check the refund, process it, update records, and inform the customer automatically. This shift—from answering questions to doing actions—turns AI into real working power.

Another mistake is expecting AI to be perfect in one response. Humans don't work that way, and AI agents shouldn't either. Good agents work step by step. They understand the request, find missing details, plan actions, try one step, check the result, and then continue or fix mistakes. This may seem slower, but it works much better in real life.

At a large scale, success is not decided by technology alone. **Trust** matters more.

AI usually fails not because it is weak, but because people are afraid. They worry about money loss, legal problems, or not knowing who is responsible if something goes wrong. That is why AI must be built with human control from the start. Leaders need to see what the AI did, why it did it, and how to stop it if needed.

Smart teams don't try to build one AI that does everything. They start small. They pick one simple, high-volume task. They test AI in the background first, where it suggests actions and humans approve them. Only after trust is built do they allow full automation.

This builds confidence and avoids risk.

In the end, the biggest challenge is not technology—it is culture. AI fails when people feel replaced. It succeeds when people feel supported. The best teams treat AI like a digital helper that does repetitive work and frees humans to think, decide, and care.

Agentic AI is not about smarter machines.

It is about redesigning how work is done.

The winning teams will not be the ones with the best demos. They will be the ones who quietly build systems where thinking leads to action—and action leads to real results.

That is the true power of AI that acts.

Top 3 learning:

1. **Where are humans still acting as "glue" between systems in our work?**

   This question helps teams identify hidden coordination work—manual handoffs, copy-paste steps, and tool switching—that agentic AI could handle end-to-end. It shifts attention from AI features to real workflow friction.

2. **Are we measuring AI success by how good the output looks—or by whether the outcome is completed?**

   This question reframes success from response quality to goal completion. It challenges teams to move from demos and pilots toward real operational impact.

3. **What would need to be true for leaders and teams to trust an agent to act on their behalf?**

   This surfaces the real blockers to scale—fear, accountability, visibility, and control. It opens conversations about guardrails, human-in-the-loop design, and trust-building rather than model capability.

Case study 2

From Signals to Conversations: Reimagining AI in Industrial Maintenance

The following example draws on publicly discussed developments in industrial automation and AI-assisted maintenance systems. It illustrates how advanced analytics and generative AI are beginning to reshape decision-making in complex industrial environments.

When Prediction Was Not Enough

For years, major industrial technology companies have invested heavily in predictive maintenance.

Using IoT sensors, machine learning models, and industrial data platforms, factories can now detect early warning signs of equipment failure. Motors, bearings, pumps, and turbines generate continuous signals about vibration, temperature, and operational behavior.

Advanced systems analyze this data and forecast when failures are likely to occur.

The operational impact of these systems has been significant:

- Reduced unplanned downtime
- Improved maintenance scheduling
- Lower repair costs
- Greater operational visibility

However, many organizations discovered an unexpected challenge.

Even when systems predicted failures accurately, technicians on the factory floor still faced difficult decisions.

Dashboards generated alerts.

Monitoring tools flagged anomalies.

Maintenance teams knew something might fail soon.

But they often struggled to answer more practical questions:

What should we fix first?

Why is this issue more important than others?

What is the best corrective action?

Prediction had improved visibility.

It had not fully solved the **decision problem**.

A Shift in Perspective

Industrial organizations gradually realized that maintenance challenges were not purely technical.

They were also deeply human.

In many sectors:

- Experienced technicians were approaching retirement.
- Critical knowledge existed in scattered manuals, maintenance reports, and personal notes.
- Younger engineers struggled to interpret alerts without contextual experience.
- Teams faced increasing "alert fatigue" from monitoring systems.

As a result, the key question began to change.

Instead of asking:

"How can we predict failures more accurately?"

Organizations started asking:

"How can we help people understand, interpret, and act on those predictions more effectively?"

This shift led to experiments with integrating **generative AI and conversational interfaces** into industrial maintenance systems.

The Evolution of AI in Maintenance

Across many industrial organizations, AI capabilities in maintenance have evolved through several stages.

1. Predictive Systems: Detecting Future Failures

In the early stage, AI analyzed sensor data to forecast potential equipment failures.

These systems generated probability scores and alerts based on operational patterns.

Value created:

Early detection of mechanical problems.

Limitation:

The information often appeared as charts and technical signals that required expert interpretation.

2. Prescriptive Systems: Recommending Actions

The next stage introduced rule-based guidance.

Maintenance systems began linking predictions to recommended actions—such as replacing a component, adjusting lubrication schedules, or planning downtime.

Value created:

Clearer guidance for technicians.

Limitation:

Recommendations were still relatively rigid and system-driven.

3. Collaborative Systems: Conversational Decision Support

More recent developments incorporate generative AI and natural-language interfaces.

In these systems, technicians can interact with maintenance intelligence through conversational queries.

For example, a technician might ask:

"Show me similar bearing failures and suggest a maintenance plan."

The system can then analyze multiple information sources, such as:

- historical sensor data
- previous maintenance records
- equipment manuals
- archived engineering notes

The AI summarizes findings in plain language and proposes possible next steps.

Instead of static dashboards, the system becomes an **interactive assistant** supporting decision-making.

Key Lessons from the Transformation

Several important insights have emerged from organizations experimenting with this new generation of AI-enabled maintenance systems.

1. Context Matters More Than Raw Data

Sensor data alone rarely provides the full picture.

The most valuable insights often come from combining operational data with human knowledge—such as technician notes, maintenance logs, and engineering documentation.

When AI systems can analyze both structured data and human experience, their usefulness increases significantly.

2. Managing Attention Is as Important as Detecting Problems

Early monitoring systems generated large volumes of alerts.

While technically accurate, the volume of signals often overwhelmed teams.

Newer systems increasingly focus on prioritization—grouping related issues and highlighting the most critical problems based on operational impact.

The objective shifts from generating more alerts to helping teams **focus attention where it matters most**.

3. AI Can Help Preserve Expertise

Industrial sectors face growing skills gaps as experienced technicians retire.

Conversational AI systems can capture and organize expert reasoning embedded in documentation and historical records.

This allows less-experienced workers to access guidance that previously existed only in the memories of senior specialists.

Rather than replacing expertise, the technology can help **scale and preserve it**.

4. Human-Centered Design Drives Adoption

Interestingly, one of the most important adoption factors has not been the sophistication of the algorithms.

It has been the **user interface**.

When AI systems communicate through complex dashboards and technical charts, adoption tends to be limited.

When they communicate through natural conversation and clear explanations, adoption improves dramatically.

In practical work environments, accessibility often determines whether AI tools become truly useful.

Turning Strategy into Operational Reality

Organizations implementing these systems typically combine several strategic steps:

Integrated Data Infrastructure

Connecting operational technology systems with enterprise data platforms to eliminate information silos.

Real-World Piloting

Testing AI systems in live industrial environments before scaling deployment.

Workforce Training

Helping employees learn how to interpret AI insights rather than blindly following recommendations.

Human Oversight

Maintaining clear governance structures so that critical decisions remain under human control.

In these environments, AI autonomy may increase over time, but trust remains a central priority.

A New Way to Think About Maintenance

These developments suggest a new way to understand predictive maintenance.

Rather than focusing only on sensors and algorithms, organizations are beginning to design systems where machines, AI, and humans interact more fluidly.

One way to describe this shift is:

The machine generates signals.

AI interprets the signals.

Humans make the decisions.

Industrial systems become not just collections of equipment, but **learning environments where data, experience, and technology interact continuously**.

Why This Matters

The broader lesson extends far beyond maintenance.

AI transformation is rarely just a technology upgrade.

It is a redesign of how **knowledge flows between people, machines, and organizations**.

When AI systems support understanding, collaboration, and decision-making, they move from being analytical tools to becoming genuine partners in complex work environments.

The future of industrial AI may therefore depend less on prediction accuracy and more on **how effectively humans and intelligent systems learn together**.

Interpreting the Case Through a Leadership Lens

Viewed through a leadership and organizational capability perspective, several themes emerge from these developments.

Successful AI adoption appears to depend not only on technical performance but also on:

- the ability to interpret complex signals
- building trust between humans and intelligent systems
- accelerating organizational learning
- maintaining human oversight in decision-making
- designing workflows that integrate AI naturally into daily work

In this sense, AI transformation becomes less about deploying algorithms and more about **designing environments where human and machine intelligence evolve together**.

Note: This case discussion draws on publicly described developments in industrial AI systems and is presented for analytical and educational purposes. It does not represent internal operations or proprietary information from any specific organization.

Case study 3

Case Illustration: Transforming Organizational Learning with AI

The following case is a composite scenario inspired by patterns observed across several telecommunications and service organizations adopting AI-enabled performance systems. Names, timelines, and operational details have been adapted for illustrative purposes.

The Hidden Contradiction

When the executive committee of a large telecommunications provider—here referred to as **TelNova**—reviewed its annual performance dashboard, an uncomfortable contradiction surfaced.

The organization had never had more data.

Millions of customer calls were recorded and analyzed.

Field service systems logged thousands of technician interactions daily.

AI-based monitoring tools tracked performance metrics and flagged operational anomalies.

Yet overall performance had plateaued.

Customer calls were still transferred too frequently.

Field technicians were repeating visits to resolve the same issues.

Training budgets continued to rise, but capability growth across the workforce remained inconsistent.

Employees described the learning ecosystem as overwhelming. Thousands of training modules existed across multiple platforms, yet frontline workers struggled to find the knowledge they needed in real time.

One senior executive summarized the situation simply:

"We don't have a data problem.

We have a digestion problem."

That realization triggered the initiative later known internally as **Project PRISM**.

Seeing the System Before Fixing It

Instead of launching another technology deployment, the organization began with something unusual.

The leadership team paused for several weeks to understand the system.

During this period, the focus was not on building solutions but on **observing how work actually happened**.

Call transcripts were analyzed.

Field service logs were mapped.

Customer satisfaction scores were compared against operational patterns.

Capability maturity was assessed across different job roles.

But the most important insight did not emerge from dashboards.

Leaders and transformation specialists spent time alongside call-center agents and field technicians. They listened to conversations, watched workflows, and asked difficult questions.

Agents explained that they were overwhelmed by training content.

Thousands of generic learning modules existed, yet very few addressed the exact problems they faced with customers.

Field technicians described similar challenges.

Recurring technical issues occurred frequently, but quick contextual guidance was rarely available at the moment it was needed.

Over time, the organization recognized a fundamental insight:

The challenge was not effort.

The challenge was **relevance**.

From Learning Repository to Intelligence Engine

Project PRISM was eventually designed not as a traditional training platform, but as a **capability intelligence system integrated into daily workflows**.

The concept was simple: instead of offering large libraries of generic training content, the system would analyze operational signals and deliver targeted learning interventions at the moment they were needed.

Using AI models trained on operational data, the system identified patterns in individual performance.

For example, if a call-center agent encountered repeated difficulty activating a digital SIM for customers, the system could trigger a brief learning intervention within the workflow—often lasting only a few minutes.

Rather than a long training course, the system delivered **contextual reinforcement tied to the exact friction point**.

Later, supervisors could review and validate these recommendations, reinforcing both learning and trust in the system.

Field technicians experienced similar support.

If a technician's upcoming schedule included a recurring network fault, the system might provide a short audio briefing summarizing the most common causes and recommended diagnostic steps before the visit.

Learning stopped being a separate event.

It became integrated into the rhythm of work.

Turning Feedback into Measurable Insight

Another important feature of the system was transparency.

Traditionally, performance feedback relied heavily on managerial interpretation. The new system introduced dashboards—often referred to internally as **capability cockpits**—that visualized learning progress and skill development patterns.

These dashboards did not replace managers.

Instead, they augmented managerial judgment.

Supervisors reviewed AI-generated recommendations and validated their relevance before deploying them widely. Over time, managers shifted from assigning generic training content to **coaching capability development more strategically**.

The system also evaluated the effectiveness of learning interventions.

If a particular micro-intervention failed to improve first-time resolution rates, it was adjusted or replaced. Learning pathways evolved dynamically as new data emerged.

Feedback gradually became more measurable and evidence-based, while still maintaining human oversight.

A Cultural Shift

Over time, subtle cultural changes began to appear.

Call-center agents started reviewing their dashboards voluntarily—not because they were required to, but because the information helped them improve performance.

Field managers began discussing capability development during operational reviews.

Instead of asking, *"Why did this problem occur?"*

They increasingly asked, ***"What capability should we strengthen next?"***

One of the most interesting patterns was that **average performers improved the fastest**. Personalized learning signals helped individuals identify development opportunities that were previously hidden inside generic training systems.

Gradually, the organization moved from episodic improvement toward a more adaptive learning culture.

Observed Operational Outcomes

Within a year of implementing the system, several measurable changes were observed:

- First-time resolution rates improved
- Call transfers decreased
- Repeat field visits declined
- Training engagement increased significantly

More importantly, employees reported greater confidence in handling complex interactions.

Managers noted fewer escalations and smoother problem resolution.

Customers experienced more consistent service journeys.

Yet the most important transformation was not purely operational.

It was psychological.

Initial concerns that AI systems might monitor or replace employees gradually shifted toward a different understanding.

The technology was not replacing expertise.

It was **augmenting human capability**.

What the Organization Actually Built

In hindsight, the initiative was not primarily about deploying artificial intelligence.

It was about redesigning how intelligence flows through the organization.

Data was interpreted before action was taken.

Learning adapts continuously.

Human oversight remained central.

Workflows were redesigned to support capability development.

Rather than treating training as a separate activity, the company embedded learning directly into operational systems.

The result was a shift from static training programs to what might be described as **a living capability architecture**.

The Broader Lesson

The experience illustrates an important principle for organizations adopting AI.

Competitive advantage rarely comes from algorithms alone.

It emerges when organizations design systems where **human expertise and machine intelligence learn together**.

In such environments, AI does not simply automate processes.

It becomes a catalyst for continuous learning and capability growth.

Note: This case example is a composite scenario created for educational purposes. It draws on patterns observed across multiple organizations exploring AI-enabled performance systems and does not represent the internal operations of any single company.

Case study 4

Case Example: AI-Driven Epigenetic Discovery Platforms

The following case example is based on publicly available information, conference presentations, and company announcements describing advances in AI-enabled epigenetic research. It is presented here for educational and analytical purposes to illustrate how artificial intelligence, and biological science can combine to accelerate discovery.

A biotechnology company working in the emerging field of **AI-enabled epigenetic medicine** has developed platforms that integrate machine learning with genomic and cellular data to explore new therapeutic pathways for complex diseases such as obesity, cardiometabolic disorders, and liver disease.

Epigenetics refers to the chemical marks that regulate how genes are activated or suppressed in different cells. Understanding these patterns is extremely complex because gene regulation depends on interactions across thousands of biological variables.

Recent advances in artificial intelligence and large-scale biological datasets are enabling researchers to analyze these patterns at a much deeper level.

AI-Enabled Epigenomic Mapping

One component of the company's platform focuses on analyzing epigenetic signals across the genome using AI-assisted computational models.

These models examine patterns of chemical modifications—such as DNA methylation—across large numbers of cells. By comparing epigenetic patterns between healthy and diseased tissues, researchers can identify regulatory signals that may contribute to disease development.

Such approaches allow scientists to study biological systems at the **single-cell and genome-wide scale**, which was previously extremely difficult using traditional laboratory methods alone.

The goal is to move beyond simply identifying correlations and instead better understand **causal mechanisms in gene regulation**.

Precision Epigenetic Engineering

Another reported capability involves using computational predictions to guide targeted modification of epigenetic marks.

In contrast to conventional gene-editing approaches that alter DNA sequences, epigenetic editing focuses on modifying regulatory signals that influence how genes behave.

According to research presentations, the platform can coordinate multiple epigenetic adjustments simultaneously with relatively high efficiency, potentially allowing researchers to influence several biological pathways at once.

Because these modifications do not necessarily require cutting the DNA sequence itself, the approach may reduce certain technical risks associated with traditional gene-editing methods.

Such capabilities could expand possibilities for studying and eventually treating complex diseases where many biological pathways interact.

Scientific Research and Early Findings

Several conference presentations have described early research results generated through this type of AI-enabled epigenetic platform.

For example:

Epigenetic Mapping of Adipocytes

At a major metabolic disease research conference, scientists presented a genome-wide map of epigenetic patterns in human fat cells. The study identified differences in epigenetic signatures between individuals with varying body mass indices, highlighting potential regulatory mechanisms involved in metabolic disorders.

Epigenetic Signals in Liver Disease

Another scientific meeting highlighted research examining epigenetic patterns in liver and adipose cells. The findings suggested that non-coding regulatory changes may influence pathways involved in liver disease development.

Multiplex Epigenetic Engineering

Additional research presentations described experiments demonstrating simultaneous modification of multiple epigenetic sites within a single system. Such multiplex approaches could allow researchers to study complex biological interactions more efficiently.

These studies illustrate how combining biological datasets with AI-driven analysis can reveal patterns that may be difficult to detect using conventional techniques alone.

Scientific and Organizational Foundations

Developing platforms that combine advanced biology, and artificial intelligence requires expertise across several domains.

According to public reports, the company was founded by scientists and technologists working at the intersection of genomics, computational biology, and machine learning. The organization also received substantial venture funding to support the development of its research platform and potential therapeutic programs.

The convergence of **biological science, data engineering, and AI modeling** represents a growing trend in modern biotechnology.

Why AI-Enabled Epigenetics Matters

The emergence of AI-driven epigenetic platforms highlights several broader shifts occurring in biomedical research:

1. AI Accelerates Biological Discovery

Machine learning systems can analyze extremely large biological datasets and identify patterns that would be difficult for human researchers to detect manually.

2. Complex Diseases Require Multi-Pathway Approaches

Conditions such as obesity, metabolic disorders, and liver disease involve numerous interacting biological mechanisms. Epigenetic analysis can reveal regulatory networks influencing these diseases.

3. Continuous Learning Loops in Scientific Research

As experimental data improves computational models, those improved models generate better predictions, which in turn guide new experiments. This creates a **self-reinforcing learning cycle**.

In this sense, AI does not merely accelerate existing research processes—it can enable entirely new forms of biological exploration.

Interpreting the Case Through a Leadership Lens

The development of such platforms also illustrates several organizational capabilities required for successful AI-driven innovation.

Rather than focusing only on algorithms, these initiatives require coordinated intelligence across scientific research, data systems, leadership decision-making, and organizational learning.

The approach described above can be interpreted through a leadership capability lens that emphasizes several key principles.

Sensemaking in Complex Systems

Biological systems involve enormous complexity. Genes, regulatory signals, and environmental influences interact across multiple layers.

Organizations working in this field must interpret vast amounts of data and distinguish meaningful biological signals from noise.

Strong sensemaking capabilities help teams:

- clarify the real scientific questions
- align researchers across disciplines
- avoid misinterpreting patterns that may not represent causal relationships

Without disciplined interpretation, AI models may identify correlations without deeper biological understanding.

Trust and Collaboration Across Disciplines

AI-driven biotechnology requires collaboration between biologists, data scientists, engineers, and clinical researchers.

Such environments demand strong communication and psychological safety so that experts from different fields can challenge each other's assumptions constructively.

When teams operate with trust and openness, complex innovation becomes more stable and resilient.

Continuous Learning and Adaptation

Fields such as epigenetics and artificial intelligence evolve rapidly. New datasets, computational tools, and experimental techniques appear frequently.

Organizations that succeed in this environment typically build **continuous learning systems**, where:

- experiments generate new data
- data improves computational models
- improved models guide further experiments

Over time, the organization's ability to learn becomes its most important advantage.

Building Reusable Scientific Platforms

Another important capability is designing systems that extend beyond a single experiment or research program.

Instead of developing isolated AI models, successful organizations often build **platform architectures** that can be applied to multiple research questions.

This allows scientific insights, datasets, and algorithms to accumulate and compound over time.

Responsible Innovation and Governance

Biotechnology innovations can have significant implications for human health. As a result, strong governance and ethical oversight are essential.

Responsible innovation requires:

- careful scientific validation
- transparent reporting of results
- clear accountability for research decisions

In high-impact fields, technological power must be balanced with thoughtful leadership.

Human–AI Partnership in Scientific Discovery

Finally, AI functions most effectively as a partner to human expertise rather than a replacement for it.

Machine learning systems can process enormous datasets and generate hypotheses at scale. Human scientists, however, remain essential for interpreting results, designing experiments, and making strategic decisions.

When human insight and machine intelligence work together, the result can be a powerful form of augmented discovery.

Key Insight

This case illustrates an important lesson about AI-driven innovation.

Breakthrough progress rarely comes from algorithms alone.

It emerges when organizations combine:

- deep domain expertise
- advanced computational systems
- disciplined leadership
- continuous learning
- responsible governance

When these elements align, AI becomes more than a tool—it becomes part of a broader system that accelerates discovery and expands human capability.

Note: This case example is based on publicly available reports and scientific presentations. It is included for analytical and educational purposes and does not imply endorsement or affiliation with the organizations mentioned.

Case study 5

You Cannot Automate Chaos: Why Process Mapping Comes Before AI

As organizations rush to adopt AI, one reality is becoming increasingly clear:

AI does not improve workflows automatically. It amplifies whatever structure already exists.

If the workflow is clean, AI accelerates it. If the workflow is messy, AI magnifies the confusion.

This is why process mapping has quietly become one of the most strategic skills in the AI era.

AI workflows are different from traditional workflows. They introduce automation layers, decision-support systems, predictive insights, and data dependencies. They blur the boundaries between human judgment and machine recommendation. They compress time. They accelerate feedback loops. And they increase the consequences of poor design.

Before AI can be embedded into any system, someone must understand how work truly flows — where it begins, how decisions are made, where delays occur, who owns accountability, and what happens when something fails.

That "someone" increasingly needs to be a leader who understands process deeply.

Process Mapping Is Not Drawing Diagrams

Process mapping is often mistaken for creating visual charts filled with boxes and arrows. But in reality, it is the discipline of seeing clearly.

It is the ability to understand how work actually moves — not how it is described in policy documents, but how it behaves in practice.

It means identifying dependencies between teams. Spotting bottlenecks that quietly slow performance. Detecting waste that drains energy. Clarifying decision points that lack ownership. And redesigning flows to improve efficiency and resilience.

In the AI era, this discipline becomes foundational.

You cannot automate chaos. You must first understand it.

The Capabilities Behind Strong Process Mapping

Strong process mapping begins with systems thinking.

A skilled practitioner does not look at isolated tasks. They see inputs and outputs, feedback loops, handoffs, and failure points. They understand that each process sits inside a larger system of incentives, tools, and behaviors. Without this systemic lens, maps become superficial.

Clarity follows systems thinking. A capable process leader asks simple but powerful questions: Where does this begin? What triggers it? Who owns each step? Where do responsibilities shift? What happens when it breaks? These questions uncover the hidden architecture of work.

Listening is equally essential. Most processes do not live in formal documentation. They live in conversations, workarounds, and informal agreements. The strongest process mappers interview stakeholders neutrally, avoid assumptions, and surface the gaps between what is written and what actually happens.

Over time, patterns emerge. Approval layers that add little value. Rework loops that consume energy. Manual duplication that exhausts teams. Misaligned incentives that distort flow. Recognizing these patterns is where mapping turns into transformation.

Communication then becomes the bridge between insight and change. Complexity must be translated into clarity. Visualizations must simplify without oversimplifying. Explanations must align teams rather than assign blame. Clarity builds trust — and trust enables redesign.

When Process Mapping Meets AI: A Case in Incident Management

Consider a mid-size SaaS company struggling with frequent production incidents. Engineers were overloaded. Alerts triggered late at night. Log analysis was manual. Slack threads were chaotic. Root causes were rediscovered repeatedly. Postmortems varied in quality. Average resolution time stretched to ninety minutes, with a significant portion of incidents recurring.

Leadership proposed a solution: "Let's use AI."

But instead of immediately deploying tools, the program manager first mapped the existing process. They traced the journey from alert to resolution to documentation. They observed repetitive manual log searches, fragmented communication, and the absence of structured knowledge reuse.

The true bottleneck was not decision-making. It was pattern recognition.

Only after understanding the system did the redesign begin.

The new workflow was introduced to AI carefully. When an alert triggers, AI now scans logs instantly. It surfaces similar past incidents, proposes likely root causes, and suggests relevant runbook steps. The engineer reviews and validates these suggestions before applying a fix. AI drafts a structured postmortem, which the engineer refines and finalizes.

The transformation was not about replacing engineers. It was about augmenting them.

AI handles scale and pattern detection. Humans retain judgment and accountability.

Within months, the mean time to resolution dropped significantly. Recurring incidents declined. Documentation improved. Cognitive load decreased.

The breakthrough did not come from AI alone.

It came from a structured redesign.

The Strategic Lesson

In the AI era, process mapping is no longer operational hygiene. It is strategic architecture.

AI compresses timelines and increases impact. Poorly designed processes become more dangerous when automated. But well-designed processes become powerful when augmented.

Exceptional process leaders can see hidden dependencies, detect systemic risk, translate complexity into structure, and balance automation with human judgment. They understand that technology should amplify clarity, not compensate for its absence.

Organizations that succeed with AI will not simply deploy smarter tools.

They will design smarter systems.

And at the center of that shift stands a leader who knows how to map, understand, and redesign the flow of work before intelligence is embedded into it.

In a world accelerating with AI, clarity is no longer optional. It is a competitive advantage.

When you redesign a workflow with AI, you need:

- Clear understanding (**Sensemaking**)
- Team trust (**Emotional Intelligence**)
- Continuous improvement (**Neuro-Learning**)
- Embedded habits (**Systemic Behavior**)
- Responsible decisions (**Ethical Intelligence**)
- Human–AI balance (^{2} **Synergy**)

AI-SENSE^2 supports every stage of the AI process integration.

Case study 6

Why Healthcare AI Fails — And How AI-SENSE^2 Prevents It?

Call centers were overwhelmed with **repetitive questions** about appointments, insurance eligibility, medication instructions, and claims status.

Staff had to manually search across multiple disconnected systems to answer even simple queries. What should have taken minutes often stretched longer, creating backlogs and frustration.

Revenue cycles were slower than they should have been. Prior authorizations required repeated follow-ups.

Claims processing involved manual checks and rework. Delays in reimbursement created financial pressure, even though the work itself was technically being completed.

Clinical and administrative teams were stretched thin. Nurses spent valuable time answering logistical questions instead of focusing on patient care. Administrative staff handled emotionally charged conversations daily, leading to fatigue and rising burnout.

Patient **experience** was inconsistent. The quality of information depended on who answered the call and how busy the system was at that moment. Growth only threatened to magnify these weaknesses.

The organization wasn't failing — but it was operating at its limits. And scaling further without structural change would have increased risk, cost, and human exhaustion.

On the surface, operations looked functional. Calls were being answered. Claims were being processed. Appointments were being scheduled. Compliance reports were being filed.

After recognizing the pressure points, the organization did not immediately launch a large AI transformation.

Instead, they started small.

First, they identified the **most repetitive and low-risk workflows** — appointment confirmations, insurance eligibility checks, and basic claims status inquiries. These were high-volume but structured tasks. They ran a limited pilot in one department, allowing AI to handle only clearly defined scenarios while humans monitored every response.

Next, they introduced **assisted workflows rather than full automation**. The AI drafted responses and surfaced relevant information, but staff reviewed and approved before sending. This helped teams build trust without surrendering control.

They also created a feedback loop. Every flagged conversation, error, or edge case was logged and reviewed weekly. Patterns were studied before expanding the scope.

Training sessions were conducted to clarify roles: AI would reduce workload, not replace judgment. Staff were encouraged to question outputs openly.

Only after stability, accuracy, and staff confidence improved did the organization gradually expand to more departments.

The transformation began not with scale — but with disciplined experimentation.

The Core Truth

The performance metrics (5× conversations, cost savings, improved outcomes) do **not** come from the model alone.

They come from:

• Workflow redesign • Governance discipline • Trust building • Continuous learning • Ethical control • Human–AI role clarity

That is exactly what AI-SENSE² structures.

AI-SENSE²: Coaching Skills for Future Leadership

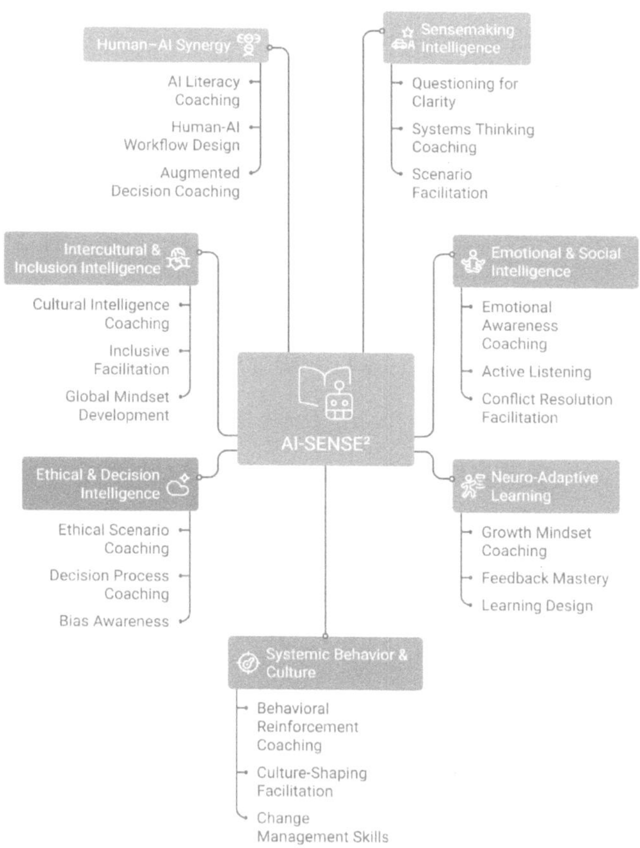

How AI-SENSE² Works Step by Step

1. S — Sensemaking Intelligence (Before Building Anything)

Most hospitals begin with the idea:

"Let's use AI to reduce call center costs."

AI-SENSE² starts differently. It first asks:

What conversations are repetitive?

Where are staff mentally overloaded?

Where are delays affecting revenue?

Which tasks require pattern recognition, and which require clinical judgment?

This step prevents blind automation.

AI should handle structured, repeatable tasks.

Humans should handle complex and unclear situations.

That is how operations can increase capacity safely—sometimes even five times more.

2. E — Emotional & Social Intelligence (During Rollout)

When AI is introduced, people feel uncertain.

Administrative staff may fear losing their jobs.

Doctors may worry about liability.

IT teams may push for speed.

AI-SENSE² requires leaders to clearly communicate:

AI supports. Humans decide.

It encourages open feedback, safe questioning of AI results, and non-punitive error reporting.

When people trust the system, they use it.

When trust is missing, usage drops, and return on investment fails.

3. N — Neuro-Adaptive Learning (After Deployment)

Many organizations launch AI once and never improve it.

AI-SENSE^2 treats deployment as the beginning, not the end.

It requires regular performance reviews, monitoring for data drift, checking for bias, tracking edge cases, and retraining models.

Because of this, accuracy improves over time.

Errors decrease.

Savings continue year after year.

4. S — Systemic Behavior & Culture (Scaling Phase)

For AI to create real impact, it must move beyond a small pilot.

AI-SENSE^2 helps organizations define decision rights, override rules, governance boards, and workflows that are fully integrated into hospital systems like EHRs.

At this stage, AI is no longer a tool.

It becomes part of daily operations.

That is how savings can scale to millions of dollars per 1,000 beds.

5. E — Ethical & Decision Intelligence (Risk Control Layer)

Healthcare is high risk.

AI-SENSE^2 ensures humans remain involved when situations are unclear.

It requires transparent audit logs, clear escalation rules, and defined accountability.

This protects the organization from legal problems, compliance issues, and reputational damage.

Trust protects long-term revenue.

6. ^{2} — Intercultural & Human–AI Synergy (Long-Term Expansion)

When expanding across regions, languages, insurance systems, and patient populations, AI must adapt.

AI-SENSE² ensures tone adjustment, cultural awareness, redesigned workflows for each region, and clear boundaries between AI tasks and human judgment.

This prevents fragmentation.

It allows global growth without losing local trust.

How Another Company Can Replicate This Success

Instead of saying, "Let's build an AI chatbot,"

The smarter sequence is:

First, map repetitive workflows.

Second, prepare staff emotionally and psychologically.

Third, build continuous learning loops.

Fourth, redesign systems before automating them.

Fifth, embed governance from the beginning.

Sixth, clearly define where AI stops and humans take over.

Only after these steps should AI agents be deployed.

Why This Produces Real Results

When AI handles structured, high-volume tasks, humans are free to focus on complex care. That is how conversation capacity increases significantly.

When manual administrative work reduces, claims move faster, rework decreases, and escalations drop. That is how cost savings grow.

When triage becomes faster, and reminders improve medication adherence, patient outcomes improve.

But these benefits only happen when leadership integrates AI properly.

What AI-SENSE² Protects Against

It protects organizations from four common failure patterns:

Over-automation

Emotional resistance

Static deployment without improvement

Unclear accountability

Final Insight

The real sequence for success is:

Leadership maturity → system design → AI deployment → continuous governance → scaled value.

When systems are aligned, automation creates an advantage.

When alignment is missing, automation creates risk.

AI-SENSE^2 ensures intelligence grows with coherence.

Case study 7

Two Engines: Code and Consciousness in the Age of AI

Imagine walking through your organization at dawn.

On one side of the building, behind glass walls and glowing dashboards, a silent engine hums. It is built of silicon and code, of data pipelines and statistical weights. This is the AI SDLC — disciplined, methodical, precise. It asks only technical questions:

Is the data clean?

Is the model validated?

Is the bias measured?

Is the deployment stable?

It lives in checklists and version control. It watches for model drift the way a pilot watches fuel levels. If something breaks, it knows which bolt to tighten, which parameter to tune. It ensures the machine is smart. It ensures the system works.

But if you keep walking, deeper into the organization — into the meeting rooms, the hallway conversations, the quiet anxieties no dashboard can detect — you will discover a second engine.

This one is not made of silicon.

It is made of belief.

This is AI-SENSE².

AI-SENSE² does not look at the code. It looks at the people holding the code. It does not monitor model drift; it monitors leadership drift. It asks a far more unsettling question:

Is our culture ready to carry this much power?

Because a perfect model dropped into a fragile culture is like installing a Ferrari engine inside a wooden carriage. The engine will roar. The wheels will shatter.

The SDLC builds the tool.

AI-SENSE² builds the hand that holds the tool.

When the algorithm recommends a radical pricing shift, the SDLC confirms the accuracy metrics. AI-SENSE² asks whether the leader has the sensemaking capacity to pause, to question, to test the ground reality beyond the dashboard.

When the model fails, the SDLC traces the bug. AI-SENSE² watches the room:

Does the team respond with curiosity — or with blame?

Does failure become learning — or fear?

The difference is subtle, but profound.

The AI SDLC manages the lifecycle of code.

AI-SENSE² manages the maturity of humans.

Think of your organization as a high-performance aircraft.

The SDLC is the engineering manual for the engine. It ensures the fuel is refined, the turbines are balanced, and the sensors are calibrated. It is the science of reliable flight.

But the manual does not teach you how to fly through a storm of human emotion.

AI-SENSE^2 sits in the cockpit.

While engineers track model drift, AI-SENSE^2 tracks cultural drift. It scans for invisible turbulence:

The quiet fear of adoption — when people feel their identity is being automated away.

Executive misinterpretation — when 90% accuracy is mistaken for certainty.

Cross-functional politics — when the data lab collides with sales reality.

Ethical tension — when quarterly pressure tempts leaders to ignore a biased output.

These are not engineering defects. They are leadership fractures.

You cannot solve identity anxiety with a better Python library.

You cannot fix ethical blind spots with a faster GPU.

The SDLC ensures the tool is sharp.

AI-SENSE^2 ensures the hand is steady.

Now imagine a massive electrical grid.

The SDLC is the insulation — CI/CD pipelines as wiring, monitoring as circuit breakers, and documentation as a blueprint. It protects the model from melting down.

But it does not manage the voltage of the organization.

AI-SENSE^2 is the transformer. It regulates the human system when the power surges.

Because AI is an amplifier.

If leadership is immature, AI will simply help you make bigger mistakes, faster.

If trust is fragile, AI will magnify suspicion.

If ethics are unclear, AI will scale the confusion.

Blind trust emerges: "The machine said so."

Defensive leadership appears: "The model made the decision."

Decision paralysis spreads as real-time data overwhelms human judgment.

Amplification without coherence creates instability.

The SDLC prevents model failure.

AI-SENSE² prevents leadership catastrophe.

Or think of it as a high-speed rail system.

The SDLC ensures the tracks align and the signals prevent collision. It operates at the system layer — mechanical, measurable, dependable.

But outside the window, there is cognitive weather: ambiguity, politics, fear, ego, urgency.

AI-SENSE² strengthens three layers simultaneously:

Self — the leader's internal clarity and regulation.

Team — the trust and emotional intelligence of the culture.

System — the habits and governance that shape decisions at scale.

A reliable machine inside an unreliable culture is not progress. It is a delayed crisis.

History shows us something deeper.

Once, AI was a furnace in the basement — analytical, contained, managed by specialists. The SDLC was sufficient because AI was a tool.

Today, AI drafts strategy, writes code, shapes customer conversations. The furnace has become the atmosphere.

We are no longer managing a tool.

We are managing intelligence embedded in the bloodstream of the enterprise.

And here is the uncomfortable truth: most modern AI failures are not mathematical errors. They are maturity failures.

Governance without discipline.

Adoption without trust.

Data without wisdom.

Accuracy without ethics.

Intelligence is scaling exponentially.

Human coherence is not.

This widening distance is the Coherence Gap.

When intelligence rises faster than maturity, confusion accelerates.

Small ethical blind spots become global liabilities.

Fear fractures culture.

Trust erodes under black-box decisions.

The AI SDLC guarantees a powerful engine.

AI-SENSE² guarantees the organization does not collapse under its own intelligence.

Imagine a racing team entering a new era.

The SDLC builds a car capable of 300 mph.

AI-SENSE² ensures the driver has the calm nervous system to handle the speed, the crew has the emotional intelligence to work under pressure, and the team remembers why the race is being run in the first place.

An engine without driver maturity is not innovation.

It is a catastrophe waiting for velocity.

In one line:

The AI SDLC ensures the AI works.

AI-SENSE² ensures the intelligence scales without the organization tearing itself apart.

In a world where machine intelligence is becoming a commodity, coherence is the true competitive advantage.

The future will not belong to the companies with the smartest models.

It will belong to the organizations whose human systems are wise enough to lead them.

Case study 8

How to Crack an AI Program Leadership Interview: Real Questions, Real Answers

The AI Program Interview That Changed Everything

Ananya walked into the final round knowing this was not a "model-building" interview.

It was an AI transformation leadership role.

She had built models before.

She had led programs before.

But today, she needed to prove she could orchestrate AI at enterprise scale.

The panel included:

- The CIO
- The Head of Data Science
- A Cloud Architect
- A Compliance Leader
- And a Commercial Business VP

This was not going to be shallow.

When the interviewer asked, **"Can you explain AI SDLC clearly?"**, Ananya did not rush into technical jargon. She took a brief pause and chose to answer in a structured, thoughtful way.

She began by clarifying that AI development is not simply traditional software development with a model added at the end. In her view, the AI lifecycle starts earlier and extends further. It begins with clearly defining the business problem and aligning it with measurable outcomes. Only after that does the work move into collecting and validating the right data, ensuring it is reliable and fit for purpose.

From there, teams experiment with features and models, testing different approaches to see what performs best. But experimentation is not enough. Models must be carefully validated, not only for accuracy but also for bias and fairness. Once validated, deployment happens through disciplined engineering practices such as CI/CD and MLOps, ensuring that models are integrated safely into production environments.

She explained that the lifecycle does not end at deployment. Continuous monitoring is essential to detect performance degradation, cost issues, or model drift. When changes in data or behavior are detected, retraining is triggered, followed by another governance review.

To make it concrete, she shared how, in her previous program, they formalized this entire lifecycle with clear stage gates. No model was allowed into production unless it passed explainability checks, bias validation, and rollback readiness testing.

As she finished, the Head of Data Science nodded. It was clear she was not describing a theory from a textbook. She had built and enforced this lifecycle in real-world programs.

When she was asked whether **she had moved AI from proof of concept to production**, Ananya smiled calmly. She answered that she had done it three times, and each time she learned the same lesson: the hardest part is rarely the model itself.

She shared a story about a demand forecasting initiative. In the experimental phase, the model achieved 92 percent accuracy in offline testing. On paper, it looked exceptional. The proof of concept impressed everyone. But once deployed into the live environment, problems began to surface. Latency increased beyond acceptable limits. Upstream data pipelines were unstable and broke frequently. Most importantly, the sales teams did not trust the forecasts, even though the numbers were statistically strong.

Instead of blaming the model, she stepped back and redesigned the entire deployment approach. She introduced formal data contracts with upstream teams to ensure data stability and accountability. She implemented performance dashboards so stakeholders could see how the model was behaving in real time. She added a human-in-the-loop override capability so business users could intervene when needed. She also organized workshops to help the sales teams understand how the model worked and how it should support their decisions.

She concluded by saying that production AI is not about achieving high accuracy in a lab. It is about reliability, transparency, and user adoption in the real world.

The CIO leaned forward slightly. It was clear she was speaking from lived experience, not theory.

When the panel **asked whether she understood governance deeply**, Ananya did not respond with a simple "yes." Instead, she offered a perspective.

She explained that governance cannot be treated as a final checkpoint added at the end of development. It must be embedded into the system from the beginning. In her previous programs, this meant establishing clear model documentation standards, so every model had traceability and accountability. It meant building bias detection pipelines directly into validation workflows rather than testing fairness only after deployment. It also required role-based access controls to ensure the right people had the right level of visibility and authority, along with automated audit logs to track model decisions and changes over time.

She went further and described setting up a Responsible AI review board that evaluated high-impact models before release. In one instance, she made the difficult decision to halt a pricing model launch because analysis revealed regional bias caused by skewed historical data. Although the model met performance targets, it posed a fairness risk.

She concluded by saying that speed without governance eventually becomes reputational risk.

The Compliance Leader quietly took notes. It was clear she understood that responsible AI is not a policy document—it is an operational discipline.

When she was **asked to explain model monitoring and drift**, Ananya *moved comfortably into technical depth. She began by clarifying the difference between data drift and concept drift. Data drift, she explained, occurs when the distribution of input data changes over time. Concept drift happens when the relationship between inputs and outputs shifts, meaning the model's learned patterns are no longer valid.*

She described how her teams used metrics such as the Population Stability Index to track distribution changes and defined clear performance degradation thresholds that would trigger investigation. Monitoring was not passive; it included shadow deployments to test new models safely alongside existing ones and automated retraining triggers when drift crossed acceptable limits.

To make it real, she shared an example. In a customer churn model, they detected drift within six weeks because buying behavior had shifted significantly during the pandemic. Since monitoring systems were already in place, they identified the issue early and retrained the model before the decline affected revenue or customer retention metrics.

When she finished, the Cloud Architect did not ask for clarification. The absence of follow-up questions signalled that her explanation had been technically solid and grounded in real-world implementation.

When the interviewer **asked whether she had managed cross-team dependencies**, Ananya gave a soft laugh and replied that this was nearly sixty percent of AI program management. The technical work matters, she explained, but coordination determines whether anything actually ships.

She described how her role required constant alignment between data engineering teams who prepared pipelines, machine learning teams building models, security and legal teams ensuring compliance, product managers defining priorities, DevOps teams handling deployment, and commercial stakeholders who ultimately depended on outcomes. Each group worked with different timelines, incentives, and constraints.

In one global AI program spanning four regions, she realized early that hidden dependencies could derail progress. To address this, she created a visual dependency heatmap that clearly mapped ownership, handoffs, and risk exposure. She instituted weekly cross-functional risk reviews where blockers were surfaced openly and resolved quickly. She also standardized release criteria so that every team worked toward the same readiness definition rather than conflicting interpretations.

She concluded by saying that AI does not fail because of algorithms alone—it fails when teams operate in silos. Her responsibility, she explained, was orchestration.

The VP smiled slightly. That was exactly the mindset they were hoping to hear.

When she was asked ***whether she could communicate trade-offs to executives****, Ananya did not mention reporting tools or dashboards. Instead, she spoke about language.*

She explained that senior leaders are not interested in technical metrics like ROC curves or model architecture diagrams. What they care about is clarity around risk, cost, speed, and business impact. Her role, she said, is to translate technical complexity into strategic implications.

She described situations where she had presented trade-offs clearly: balancing model accuracy against system latency when faster decisions were more valuable than marginal performance gains; weighing infrastructure cost against retraining frequency to maintain optimal performance without overspending; and discussing how accelerating innovation might increase compliance exposure if governance controls were not mature.

In one critical case, she advised delaying a product launch. Although the model performed well, the observability framework was not yet robust enough to detect issues at scale. Releasing early would have introduced technical debt and operational risk that the organization was not ready to absorb.

As she finished, the CIO nodded thoughtfully. She had demonstrated that she understood how to frame AI decisions in terms that resonate at the board level, where consequences matter more than technical elegance.

When **the panel asked whether she had built reusable AI platforms**, Ananya answered confidently that she had. She explained that early in her *career, she had seen teams repeatedly build isolated, one-off solutions. Each project required new pipelines, new governance reviews, and new deployment setups. It was inefficient and impossible to scale.*

So, she shifted the approach from project delivery to platform thinking. She led the creation of shared feature stores so teams could reuse validated data features instead of recreating them. She introduced a standardized model registry to ensure traceability, version control, and consistency across deployments. She implemented reusable CI/CD pipelines tailored for AI workflows, allowing models to move from experimentation to production through a consistent, automated path. Governance templates were pre-approved and embedded into the lifecycle, reducing friction while maintaining compliance. All of this was supported by a cloud-native reference architecture designed for scalability and resilience.

She concluded by saying that real AI maturity begins when teams stop rebuilding the same infrastructure and start building shared capabilities that compound over time.

The atmosphere in the room subtly shifted. It became clear that they were not speaking with someone who merely managed timelines. They were speaking with someone who designs systems that enable long-term organizational capability.

When the final question came—whether she could run AI programs at scale—Ananya paused for a moment before answering. She knew this was not a question about volume. It was a question about depth.

She explained that scale is not achieved by deploying more models or launching more pilots. True scale, she said, comes from organizational maturity. It requires governance to be embedded from the beginning rather than added later. It requires automation to be the default approach so that reliability and repeatability are built into the system. It demands strong cross-functional alignment so that data, engineering, security, legal, and business teams move together rather than in fragments.

She spoke about the importance of continuous learning loops, where feedback from production informs improvement instead of being ignored. She emphasized responsible AI oversight, ensuring that ethics and compliance grow alongside technical capability. And she highlighted the need for clear executive visibility so leaders can make informed decisions without surprises.

She concluded by saying that her goal is not simply to deliver AI projects. Her goal is to elevate AI capability maturity across the organization so that intelligence becomes a sustainable, scalable strength.

The room fell quiet for a moment.

Then the CIO looked at her and asked, calmly and directly, "When can you start?"

Case study 9

Personalization at Scale: How Streaming Platforms Use AI to Create Intelligent Viewing Experiences

The Context: When Choice Becomes Overload

In the early days of digital streaming, the biggest challenge was **content availability**. Platforms competed to offer more movies and shows than traditional television.

But as streaming libraries expanded into **tens of thousands of titles**, a new problem emerged.

Too much choice.

Users were spending more time **searching for something to watch** than actually watching. Many opened the platform, scrolled endlessly, and eventually abandoned the session.

For streaming companies, this was a serious business risk.

When viewers cannot quickly find relevant content:

- engagement drops
- watch time declines
- subscriptions churn

What began as a **content distribution challenge** became a **decision-making challenge**.

The question was no longer:

"How do we provide more content?"

The real question became:

"How do we help each viewer discover the right content at the right moment?"

This shift led to one of the most influential uses of AI in modern digital products: **personalized recommendation systems**.

The Challenge: Millions of Viewers, Millions of Preferences

Streaming platforms operate at an enormous scale.

Millions of users watch content simultaneously across different devices, time zones, cultures, and moods.

Every viewer behaves differently.

Some binge entire seasons overnight.

Others watch casually on weekends.

Some prefer documentaries; others prefer short comedy shows.

Even the same user behaves differently at different times.

A viewer might watch a thriller on Friday night and a children's animation on Sunday morning.

Traditional recommendation methods—such as **editor-curated lists or static genre categories**—could not keep up with this complexity.

Platforms needed a system capable of learning continuously from:

- viewing history
- browsing behavior
- device usage
- time of day

- regional preferences
- interactions such as pause, skip, or replay

The challenge was not simply predicting what users like.

The real challenge was **understanding human behavior dynamically**.

The AI Solution: Personalization as a Learning System

Modern streaming platforms use AI to transform viewing data into personalized experiences.

Instead of presenting the same homepage to every user, the platform generates **a unique content interface for each viewer**.

This system typically works through several layers of intelligence.

Behavioral Data Analysis

AI models analyze patterns such as:

- what users watch
- how long they watch
- what they abandon early
- what they search for
- what they watch repeatedly

These signals provide insights into **viewer interests and engagement levels**.

Collaborative Filtering

Recommendation algorithms identify patterns across users.

If many viewers with similar viewing behavior enjoy a particular show, the system may recommend that show to others with similar patterns.

Content Understanding

AI systems also analyze the **characteristics of the content itself**, such as:

- genre
- themes
- pacing

- actors
- tone

By understanding both **user behavior and content characteristics**, the system can match viewers with relevant titles.

Continuous Learning

The system constantly learns.

Every click, pause, rating, or skipped episode becomes new data.

Over time, recommendations improve as the system adapts to changing viewer habits.

How This Reflects AI-SENSE² Leadership Principles

The personalization system illustrates several important principles about how intelligent systems and organizations evolve together.

Below is how the story reflects each element of the framework conceptually.

Sensemaking Intelligence

Turning Data into Meaning

Streaming platforms generate enormous amounts of behavioral data.

But data alone does not create insight.

The real value comes from **interpreting patterns in user behavior**.

Instead of asking:

"What content do we have?"

The organization began asking:

"What signals reveal what people actually enjoy?"

This shift required leaders and product teams to **see patterns hidden within complexity**.

AI provided the computational capability to detect these signals, while human product teams interpreted them within the broader context of user experience.

Sensemaking transformed raw data into meaningful recommendations.

Emotional and Social Intelligence

Understanding Human Preferences

Entertainment is deeply emotional.

People watch content not only for information but also for relaxation, excitement, curiosity, and connection.

Personalization systems, therefore, must consider **human context**, not just viewing statistics.

For example:

- weekend viewing habits differ from weekday viewing habits
- Family viewing differs from individual viewing
- mood influences content choices

Product teams designing these systems must remain aware that behind every data point is **a human moment**.

Successful personalization systems combine technical intelligence with empathy for user behavior.

Neuro-Adaptive Learning

Continuous Learning Loops

A key strength of modern recommendation systems is their ability to **adapt continuously**.

Instead of relying on static rules, AI models learn from ongoing interactions.

If a viewer begins exploring a new genre, recommendations adjust quickly.

If a user abandons a show halfway through, the system learns that the recommendation may have been inaccurate.

This creates a feedback loop:

user behavior → model learning → improved recommendations → new behavior signals

Over time, the system becomes increasingly aligned with individual preferences.

Organizations that design such systems effectively are not simply building software.

They are building **adaptive learning environments**.

Systemic Intelligence

Designing the Whole Ecosystem

Personalization does not operate in isolation.

It requires integration across multiple systems:

- content metadata
- user profiles
- recommendation models
- streaming infrastructure
- user interface design

The recommendation engine must interact seamlessly with the product experience.

For example:

- homepage rows
- suggested titles
- auto-play previews
- search results

Each of these components contributes to how users discover content.

Successful AI products emerge when organizations treat AI not as a standalone feature but as **a system integrated into the product architecture**.

Ethical and Decision Intelligence

Balancing Engagement and Responsibility

Recommendation systems influence what millions of people watch.

This creates important ethical considerations.

Algorithms optimized purely for engagement could unintentionally:

- create filter bubbles
- promote repetitive viewing patterns
- amplify certain types of content disproportionately

Responsible AI design therefore, requires **human oversight**.

Product leaders must balance multiple goals:

- user satisfaction
- content diversity
- long-term engagement
- responsible digital experiences

The role of leadership becomes critical in defining the boundaries within which AI systems operate.

Human–AI Synergy

Intelligence as Partnership

Personalization systems do not replace human decision-making.

Instead, they augment it.

AI can analyze millions of behavioral signals simultaneously—something humans cannot do.

But humans still provide essential inputs:

- product design decisions
- content strategy
- algorithm governance
- ethical boundaries

The most effective streaming platforms combine **machine intelligence and human judgment** to shape the viewing experience.

AI suggests possibilities.

Humans define direction.

The Business Impact

Personalization systems have become one of the most powerful drivers of engagement in streaming platforms.

By helping users discover relevant content quickly, AI systems can:

- increase watch time
- improve customer satisfaction
- reduce subscription churn
- support content discovery across large catalogues

Instead of overwhelming users with thousands of titles, the platform guides them toward content aligned with their preferences.

The result is not just better technology.

It is a **better decision environment for users**.

The Strategic Lesson

This case demonstrates an important insight about AI in modern organizations.

The true value of AI does not come from algorithms alone.

It comes from **how organizations design systems that learn from human behavior and adapt continuously**.

Personalization systems succeed because they combine:

- data signals
- adaptive learning
- human-centered product design
- ethical oversight

Together, these elements create an intelligent ecosystem that evolves with its users.

Leadership Insight

In the age of AI, the competitive advantage of digital platforms does not lie only in technology.

It lies in their ability to:

sense complex patterns

learn continuously

design systems intelligently

and guide AI with human judgment.

When organizations align these capabilities, AI stops being a feature.

It becomes a **living intelligence embedded within the product experience**.

Case study 10

Learning on the Road: AI Feedback Loops in Autonomous Driving Systems

The Context: Driving Is One of the Most Complex Human Activities

Driving appears simple when performed by an experienced human. Yet the task involves thousands of simultaneous judgments.

Drivers must interpret:

- moving vehicles
- pedestrians and cyclists
- road signs and traffic lights
- weather conditions
- unpredictable human behavior

A human brain processes these signals intuitively through years of experience.

For autonomous driving systems, replicating this ability is extraordinarily difficult.

Unlike many digital AI applications, autonomous driving operates in **an open, real-world environment** where conditions change constantly.

Roads are not controlled laboratories.

They are dynamic systems full of uncertainty.

This reality forced automotive companies developing autonomous capabilities to rethink how AI systems learn.

Traditional software development methods were not sufficient.

The Challenge: Teaching Machines to Understand the Real World

Early autonomous driving systems relied heavily on **rules-based programming**.

Engineers attempted to encode driving behavior explicitly:

- If a pedestrian appears, slow down.
- If the traffic light turns red, stop.
- If another car approaches, maintain distance.

But real roads rarely follow clean rules.

Unexpected scenarios occur constantly:

- unusual road markings
- construction zones
- unpredictable pedestrian behavior
- unusual weather conditions

These "edge cases" are difficult to anticipate in advance.

An autonomous driving system cannot rely solely on pre-programmed instructions.

Instead, it must **learn from real-world driving experience**.

This requirement led to the development of large-scale **learning feedback loops** between vehicles, data systems, and machine learning models.

The AI Approach: Learning From Real-World Driving Data

Modern autonomous driving systems typically combine several technologies.

Sensor-Based Perception

Vehicles are equipped with sensors such as:

- cameras
- radar systems
- ultrasonic sensors

These sensors capture real-time information about the surrounding environment.

AI models then interpret this sensor data to detect objects such as:

- vehicles
- pedestrians
- traffic signals
- road boundaries

Machine Learning for Pattern Recognition

Deep learning models analyze visual data from cameras and identify patterns within the environment.

Instead of following strict rules, these models learn from examples.

Large datasets of driving situations allow the system to recognize common road scenarios.

Continuous Data Feedback

One of the most important aspects of modern autonomous systems is the **feedback loop between vehicles and AI training systems**.

Vehicles operating in real-world conditions generate large amounts of driving data.

When unusual situations occur, those scenarios can be captured and used to improve the AI models.

This creates a learning cycle:

real-world driving → data collection → model training → software updates → improved performance

Over time, the system becomes better at recognizing complex environments.

How Learning Loops Reflect AI-SENSE² Principles

Autonomous driving development illustrates several deeper principles about how AI systems and organizations evolve together.

Sensemaking Intelligence

Understanding Complex Signals

Driving environments generate enormous volumes of sensory information.

Raw data alone is not useful unless it can be interpreted meaningfully.

AI systems must convert sensor signals into a coherent understanding of the world.

For example, the system must determine:

- which objects are moving
- which objects represent potential hazards
- Which signals require immediate action

This requires advanced **sensemaking capabilities**.

Engineers and data scientists design systems that transform raw visual data into structured situational awareness.

The leadership lesson here is important.

In complex environments, organizations must move beyond collecting data and focus on **interpreting what the signals actually mean**.

Emotional and Social Intelligence

Designing for Human Trust

Autonomous driving technology affects human safety directly.

Because of this, user trust becomes critical.

Drivers must feel confident that the system supports them rather than replacing them prematurely.

Successful implementations often emphasize **human supervision and transparency**.

Drivers remain responsible for the vehicle and must stay attentive.

This approach reinforces an important leadership principle:

When AI interacts with humans in high-risk environments, trust must be built gradually through **clear roles and responsibilities**.

Technology alone cannot create trust.

Human-centered design does.

Neuro-Adaptive Learning

Continuous Improvement Through Experience

One of the defining characteristics of modern autonomous driving systems is their ability to improve through learning cycles.

Edge cases—rare or unusual driving situations—become valuable training examples.

Instead of treating these scenarios as failures, development teams treat them as **learning opportunities**.

Each new scenario contributes to improved models.

This continuous feedback process accelerates learning.

Organizations building such systems must create infrastructure capable of:

- collecting real-world data
- labelling and interpreting events
- retraining machine learning models
- distributing updates safely

This approach reflects a deeper shift.

AI development becomes an ongoing learning process rather than a one-time engineering project.

Systemic Intelligence

Building a Complete AI Ecosystem

Autonomous driving systems depend on an integrated technological ecosystem.

This includes:

- vehicle hardware and sensors
- onboard computing systems
- cloud-based data infrastructure
- machine learning pipelines
- software update mechanisms

No single component creates the capability alone.

The real strength lies in **how the system functions as a whole**.

Organizations working in this field must coordinate multiple disciplines:

- automotive engineering
- data science
- software architecture
- safety engineering

Systemic thinking becomes essential.

Ethical and Decision Intelligence

Safety as the Ultimate Constraint

Autonomous driving involves life-critical decisions.

AI systems must operate under strict safety guidelines.

Engineers must address questions such as:

- how the system prioritizes risk
- when the system alerts the driver
- when the system disengages control

Because of these risks, human oversight remains essential.

AI systems assist drivers but do not eliminate the need for human judgment.

Responsible leadership ensures that AI capabilities evolve within **clear ethical and safety boundaries**.

Human–AI Synergy

Augmenting Human Drivers

Rather than replacing human drivers completely, many modern systems focus on **driver assistance**.

AI can:

- detect hazards earlier
- maintain safe distances
- monitor surroundings continuously

Humans still provide context, situational awareness, and final control.

This partnership allows both human and machine intelligence to complement each other.

Machines provide speed, consistency, and monitoring.

Humans provide judgment, adaptability, and accountability.

The Strategic Insight

Autonomous driving illustrates a broader lesson about AI innovation.

The most powerful AI systems are not those that attempt to replace human intelligence entirely.

They are systems that create **continuous learning loops between machines, humans, and real-world environments**.

The competitive advantage comes from how quickly the system learns and improves.

Organizations that build strong feedback loops between data, models, and human oversight can evolve faster than those relying solely on static software systems.

Leadership Reflection

In complex environments, intelligence is not a fixed capability.

It is a **dynamic process of sensing, learning, and adapting**.

Autonomous driving systems demonstrate how AI can extend human perception and learning.

But they also show that human judgment, ethical responsibility, and system design remain essential.

The future of AI is not a world where machines replace humans.

It is a world where **human and machine intelligence evolve together through continuous learning systems**.

Case study 11

The Interview That Broke Rajan – And the One That Built Him

The Confidence Before the Fall

Rajan walked into the interview room with the kind of quiet confidence that only comes from doing the work.

For three years, he had lived inside AI systems.

He had built chatbots that spoke like humans, designed pipelines that processed thousands of requests, worked with embeddings, vector databases, APIs — the full stack of modern AI engineering.

He wasn't guessing his way through this.

He had earned his place.

This interview, he believed, was his next step — a Senior AI Implementation Lead role. A role with influence, ownership, and a seat at the strategy table.

He expected a fair fight.

What he didn't expect... was silence.

The First Crack

The first question was simple.

"Tell us about a project you're proud of."

Rajan leaned forward, comfortable. This was his territory.

He spoke with clarity and depth — about the architecture he had designed, the embedding models he had chosen, the way he optimized latency, the way he handled API rate limits under pressure. Every detail was precise. Every decision had logic behind it.

It was the kind of answer that would impress any engineer.

The panel nodded.

Then one interviewer asked, almost casually:

"What impact did this create for the business?"

Rajan paused.

Not because he didn't know his system.

But because, for the first time, he realized — he didn't know how to answer that question.

Not really.

The Silent Disconnect

The next question came.

"How would you implement AI across our organisation in 90 days?"

Rajan began confidently again, outlining integrations, workflows, architecture layers. It made sense in his head. It was technically sound.

But halfway through, he noticed something shift.

The panel wasn't leaning in anymore.

They weren't confused.

They were... unconvinced.

The Question That Stayed

Then came the question that stayed with him long after the interview ended.

“What happens when your AI system fails in production?”

Rajan answered the way he always had — retries, monitoring, fallback mechanisms.

The interviewer stopped him.

“I’m not asking about systems. I’m asking about decisions.”

The room went quiet.

And for the first time in years, Rajan felt something he hadn’t felt in a long time.

Uncertainty.

The Three Words That Hurt

Two days later, the email arrived.

“Strong technical candidate, but not leadership ready.”

Three words.

But they landed heavier than anything he had heard before.

Because deep down, he knew... they were right.

The Realization

Rajan hadn’t failed because he lacked knowledge.

He had failed because he answered like an engineer...

in a room that was looking for a leader.

He spoke about how systems work.

But they were evaluating how decisions are made.

How impact is created.

How leadership shows up when things don’t go as planned.

The Decision to Change

Most people, at that point, would have moved on.

Rajan didn't.

He stepped back instead.

And that decision changed everything.

Learning to Think Again

He began working with an AI expert coach.

Not to learn new tools.

But to relearn how to think.

The first session was uncomfortable.

Rajan described one of his best projects again — the architecture, the design, the technical challenges he had solved.

His coach listened quietly, then interrupted him.

"That's what you built. Now tell me what it did."

Rajan paused again.

The same pause.

But this time, he didn't rush past it.

He sat with it.

And in that silence, something shifted.

From System to Impact

He went back to the same project and saw it differently.

Not as a system... but as a solution.

Customer support response times had been eight hours.

The system he built brought it down to twenty-three minutes.

Seventy percent faster.

Hundreds of hours saved.

Human agents are freed to focus on what actually requires human judgment.

Same work.

Different meaning.

The Inner Shift

From there, the transformation accelerated.

Mock interviews became intense. His coach interrupted him, challenged his assumptions, and pushed him into uncomfortable territory. At first, Rajan resisted. He defended his answers. He tried to prove he was right.

Then slowly, he changed.

He learned to pause before answering.

To listen fully.

To respond with clarity instead of complexity.

He stopped trying to sound smart.

He started trying to be understood.

Unlearning to Lead

He realized something even harder.

Everything that had made him a strong engineer — his focus on tools, solutions, execution — was limiting him as a leader.

So, he began to unlearn.

He stopped starting with solutions.

He started with problems.

He stopped asking, "Can we build this?"

He started asking, "Should we build this — and what happens if we do?"

Thinking in Systems

He no longer saw AI as just technology.

He began to see it as leverage.

Every decision now carried weight:

What happens when these scales are used?

What breaks under pressure?

What does this cost at ten times the usage?

What value does it create — and for whom?

Speaking the Language of Business

He practiced translating.

From AI to business value.

From models to money.

From systems to strategy.

He learned to speak to engineers, yes — but also to product leaders, to operations heads, to decision-makers who didn't care about embeddings but cared deeply about outcomes.

The Second Chance

Six weeks later, Rajan walked into another interview.

Same level. Same expectations.

But this time... a different presence.

A Different Answer

When they asked about his project, he didn't start with architecture.

He started with impact.

"We were facing an eight-hour delay in customer support. I designed a system that reduced that to twenty-three minutes, improving efficiency by seventy percent and allowing our team to focus on high-value work."

This time, the panel leaned forward.

Clarity Over Complexity

When they asked about the first 90 days, he didn't rush into building.

He paused.

Then said:

"In the first 30 days, I wouldn't implement anything. I would understand the highest-cost problems and map them to available data. In the next 30, I would deliver one high-impact, low-risk win to build trust. In the final 30, I would scale that success and align the organisation for broader transformation."

No jargon.

No over-explanation.

Just clarity.

Owning Failure

When they asked what happens when AI fails, he didn't defend the system.

He acknowledged reality.

"AI will fail. The question is whether we designed for it. I build systems where failures are contained — with validation layers, human checkpoints, and clear escalation paths."

The Outcome

Four days later, the offer came.

The Real Transformation

Rajan didn't become better at AI.

He became better at thinking.

At deciding.

At communicating.

At the leading.

The Final Truth

Because in the end, AI doesn't create leaders.

It amplifies them.

If your thinking is unclear, AI scales confusion.

If your thinking is sharp, AI scales impact.

Rajan didn't just crack the interview. He became the kind of leader the role had been waiting for.

Conclusion

The Leaders Who Will Matter Most

As this book ***Leading When Machines Think Faster, "Winning with AI: The Leadership Playbook for Enterprise Advantage"*** ends, one truth stands clearer than ever:

The AI-SENSE² Leadership Map

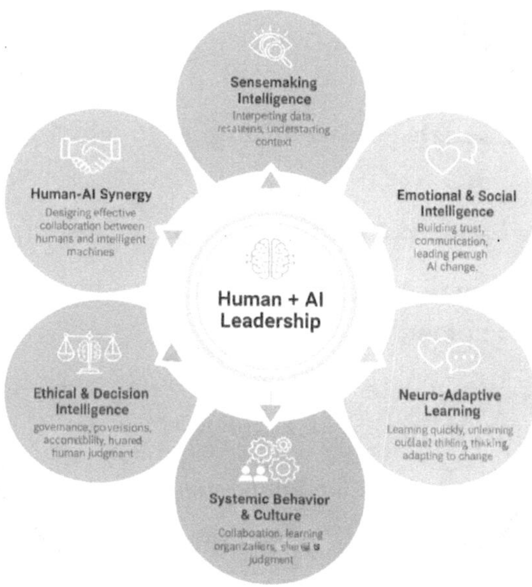

AI leadership requires multiple internocted human capabilties working together with artificial intelligence to navigate complexity and create enterprise advantage.

The defining leadership challenge of our time is not artificial intelligence—it is **human readiness**.

Machines now think faster than we ever could. They analyze patterns at scale, generate insights in seconds, and act without fatigue. Yet despite all this intelligence, something essential remains unresolved. Machines can optimize decisions, but they cannot decide what *should* matter. They can predict outcomes, but they cannot carry responsibility for those outcomes. They can move fast, but they cannot choose wisely.

That burden—and that privilege—still belongs to leaders.

Throughout this book, we have explored why leadership must evolve in the AI era. Not incrementally. Not cosmetically. But fundamentally. We have seen how traditional leadership models—built on control, expertise, speed, and certainty—begin to fracture when intelligence becomes non-human, non-linear, and ever-present.

AI did not break leadership.

AI **revealed its limits**.

What emerges in response is not a new toolset, but a new kind of leader.

A leader who can **sense before acting**, even when pressure demands speed.

A leader who can **interpret meaning**, not just consume data.

A leader who can **pause AI**, not because it failed technically, but because judgment requires humanity.

A leader who can regulate emotion, sustain trust, and lead people through uncertainty rather than around it.

A leader who learns faster than change—and helps others do the same.

A leader who understands that culture, coaching, and behavior determine whether intelligence becomes impact.

This is the leader the AI era demands.

AI-SENSE² Is Not a Framework You Apply—It Is a Way You Lead

AI-SENSE² was never intended to be another model to memorize or a checklist to implement. It is a **leadership operating system**—one that integrates sensemaking, ethics, emotional intelligence, learning agility, systemic behavior, and human–AI partnership into a coherent whole.

It exists because leadership can no longer afford fragmentation.

You cannot separate strategy from ethics.

You cannot separate intelligence from emotion.

You cannot separate speed from responsibility.

You cannot separate AI from the humans who authorize it.

AI-SENSE² holds these tensions together.

This book reminds us that leadership is not about outperforming machines—but about **guiding intelligence toward outcomes that are meaningful, just, and sustainable**.

Succeeding with AI is not mainly a technology problem.

It is a **human capability problem**.

Organizations do not struggle with AI because they lack tools, data, or models. They struggle because thinking, learning, decision-making, and leadership have not evolved at the same pace as technology.

That is exactly the gap AI-SENSE² is designed to close.

By building sensemaking, emotional intelligence, learning agility, ethical judgment, systemic behavior, intercultural awareness, and human–AI collaboration, AI-SENSE² develops the full set of skills required to design, deliver, and scale real AI solutions. These are not "soft" skills. They are **core skills for AI work**.

Global research from OECD, McKinsey, and Deloitte confirms this shift. The leaders and teams who succeed with AI are not those who know the most about algorithms, but those who can think clearly in complexity, regulate emotions under pressure, learn faster than change, shape culture intentionally, and make responsible decisions at speed.

AI-SENSE² does not train people to simply *use* AI.

It trains them to **work intelligently with AI**.

That is why the framework prepares leaders, product teams, and organizations not just to adopt AI—but to make it work in the real world, at scale, and over time.

As AI continues to evolve, technical skills will keep changing.

Human capabilities must become stronger, not optional.

The future belongs to those who can combine human judgment with machine intelligence—wisely, ethically, and continuously.

This book closes with a simple truth:

AI success is built on human strength.

AI-SENSE² builds that strength.

What comes next is not another tool or model—but a deeper way of leading, thinking, and creating value in the age of intelligent systems.

The Future Will Not Reward the Fastest Leaders—It Will Reward the Wisest

In the years ahead, organizations will deploy more advanced AI. Agentic systems will act with greater autonomy. Decisions will be made at unprecedented speed. And the temptation to step back—to let machines decide—will only grow stronger.

But history will not remember who automated the most.

It will remember who **led with clarity when it was easier to delegate judgment**.

Who slowed down when speed was dangerous?

Who chose learning over certainty?

Who protected humanity when efficiency demanded sacrifice?

The leaders who will matter most are not those who know the most about AI.

They are those who can:

- *Sense what is changing before it breaks*
- *Learn faster than their own expertise expires*
- *Decide responsibly when outcomes are irreversible*
- *Lead people through ambiguity without losing trust*
- *Work with AI without surrendering wisdom*

Three Thinkers. One Big Idea.

There is a small café at the edge of time.

No Wi-Fi. No notifications. Just three chairs, a round table, and the quiet hum of deep thought.

One evening, three visitors arrive.

An old man in a worn robe. A wild-haired physicist still scribbling on a napkin. And a modern leader, laptop open, eyes full of questions.

They are here to talk about intelligence — and what it means now that machines have some of it too.

The Old Man Orders First

Socrates wraps both hands around his cup and speaks before anyone asks him anything.

"Tell me," he says, *"what frightens you more — a machine that thinks faster than you, or a human who has forgotten how to think deeply?"*

Nobody answers straight away. Which, of course, is exactly what he wanted.

For Socrates, wisdom was never about speed. **It was about the quality of the question underneath the answer.** He spent his entire life wandering Athens not to give people answers, but to unsettle the answers they already thought they had.

In a world where AI can generate a response in milliseconds, he leans forward and says quietly:

"The machine will always answer faster. But only you can ask: is this the right question? What are we assuming? What might we regret?"

He picks up his cup.

"That is not a technical skill. That is wisdom. And wisdom cannot be downloaded."

The Physicist Looks Up from His Napkin

Einstein has been listening. He always listens before he speaks.

He pushes the napkin across the table. On it is not an equation — just two words: *Imagine. Connect.*

"Knowledge tells you what already exists," he says. *"Imagination shows you what could."*

His greatest discoveries never came from processing more data. They came from picturing a boy riding a beam of light. From wondering what the universe looks like from inside a falling elevator. From connecting ideas that had never been in the same room before.

He taps the napkin.

"Your AI systems are extraordinary at finding patterns in what has already happened. But the breakthrough — the genuinely new thing — that still comes from here." He touches his temple. *"Human imagination, amplified by machine learning. That is the most powerful combination in history."*

He pauses.

"The danger is not that machines become too intelligent. The danger is that humans become too passive."

The Modern Leader Closes the Laptop

She has been nodding since the moment she sat down.

She leads a global organisation of twelve thousand people across nineteen countries. She has deployed AI in operations, finance, and customer experience. And she will tell you honestly — the hardest part was never the technology.

"We built the model in four months," she says. *"It took us two years to get the organisation to trust it."*

She has read Socrates. She admires Einstein. But her daily reality is messier than philosophy and more complex than physics.

Her questions are different:

How do you help a team in Mumbai collaborate with a system trained on data from Chicago? How do you build a culture where people learn faster than the technology changes around them? How do you make ethical decisions at machine speed, across six time zones, in three languages?

"What we needed," she says, *"was not a better algorithm. We needed better leaders. People who could sense what was really happening, learn continuously, and guide human-machine teams with both confidence and humility."*

She looks at the other two.

"You were both right. We just needed a way to scale it."

The Bill Arrives

Three thinkers. Three centuries. Three completely different worlds.

And yet they had arrived at exactly the same place.

The future does not belong to the fastest machines.

It belongs to the humans who learn how to think *alongside* them — questioning deeply like Socrates, imagining boldly like Einstein, and building at scale like the leaders the world needs right now.

Intelligence, they agreed as they stood to leave, is no longer just something you possess.

In the age of AI, it is something we co-create.

A Final Invitation

This book does not end with answers.

It ends with an invitation.

An invitation to lead differently.

To think more deeply.

To remain human—intentionally—when machines grow more intelligent.

AI will continue to evolve. That is inevitable.

The question is whether leadership will evolve with it.

The future belongs to leaders who can **sense, learn, decide, and lead with AI—without losing their humanity**.

That future is not abstract.

It begins with how you lead—today.

References

Foundational AI & Enterprise Transformation

1. **Competing in the Age of AI**

   Iansiti, M., & Lakhani, K. R. (2020). *Competing in the Age of AI: Strategy and Leadership When Algorithms and Networks Run the World.*.

   Publisher: Harvard Business Review Press

   Edition: First Edition

2. **Human + Machine**

   Daugherty, P. R., & Wilson, H. J. (2018). *Human + Machine: Reimagining Work in the Age of AI.*

   Publisher: Harvard Business Review Press

   Edition: First Edition

3. **Prediction Machines**

   Agrawal, A., Gans, J., & Goldfarb, A. (2018). *Prediction Machines: The Simple Economics of Artificial Intelligence.*

   Publisher: Harvard Business Review Press

   Edition: First Edition

4. **AI Superpowers**

   Lee, K.-F. (2018). *AI Superpowers: China, Silicon Valley, and the New World Order.*

   Publisher: Houghton Mifflin Harcourt

   Edition: First Edition

Sensemaking, Decision-Making & Complexity

5. **Sensemaking in Organizations**

   Weick, K. E. (1995). *Sensemaking in Organizations.*

   Publisher: SAGE Publications

   Edition: First Edition

6. **The Signal and the Noise**

   Silver, N. (2012). *The Signal and the Noise: Why So Many Predictions Fail—but Some Don't.*

   Publisher: Penguin Press

   Edition: First Edition

7. **Thinking in Systems**

   Meadows, D. H. (2008). *Thinking in Systems: A Primer.*

   Publisher: Chelsea Green Publishing

   Edition: First Edition

Emotional, Social & Psychological Intelligence

8. **Emotional Intelligence**

   Goleman, D. (1995). *Emotional Intelligence: Why It Can Matter More Than IQ.*

   Publisher: Bantam Books

   Edition: First Edition

9. **Primal Leadership**

   Goleman, D., Boyatzis, R., & McKee, A. (2002). *Primal Leadership.*

   Publisher: Harvard Business School Press

   Edition: First Edition

10. **The Fearless Organization**

   Edmondson, A. (2018). *The Fearless Organization.*

   Publisher: Wiley

   Edition: First Edition

Learning Agility, Neuroplasticity & Flow

11. **Mindset**

   Dweck, C. S. (2006). *Mindset: The New Psychology of Success.*

   Publisher: Random House

   Edition: First Edition

12. **Peak**

   Ericsson, A., & Pool, R. (2016). *Peak: Secrets from the New Science of Expertise.*

   Publisher: Houghton Mifflin Harcourt

   Edition: First Edition

13. **Flow**

   Csikszentmihalyi, M. (1990). *Flow: The Psychology of Optimal Experience.*

   Publisher: Harper & Row

   Edition: First Edition

Ethics, Responsibility & AI Governance

14. **Weapons of Math Destruction**

    O'Neil, C. (2016). *Weapons of Math Destruction.*

    Publisher: Crown Publishing

    Edition: First Edition

15. **The Alignment Problem**

    Christian, B. (2020). *The Alignment Problem: Machine Learning and Human Values.*

    Publisher: W. W. Norton & Company

    Edition: First Edition

Human–AI Collaboration & Future Leadership

16. **The Algorithmic Leader**

    Walsh, M. (2019). *The Algorithmic Leader.*

    Publisher: Page Two

    Edition: First Edition

17. **Reinventing Jobs**

    Jesuthasan, R., & Boudreau, J. (2017). *Reinventing Jobs.*

    Publisher: Harvard Business Review Press

    Edition: First Edition

www.ingramcontent.com/pod-product-compliance
Lightning Source LLC
LaVergne TN
LVHW100248210926
848480LV00003B/857

* 9 7 9 8 9 0 4 5 1 1 8 3 8 *